C.M. (Dad) Joiner brought the flow of wealth

DAD
&
DOC

By

Michelle M. Haas

WITH

Harry Harter's

EAST TEXAS
OIL PARADE

Copano Bay Press
2014

ISBN 978-1-941324-07-3

Contents

An Informal Bibliography

In addition to my personal collection of Joiner and 1920s oil promoters' sales letters and ephemera, much of the source material for the biographies of Dad and Doc came from newspaper archives spanning the country. The bulk of periodical material used was drawn from the *Dallas Morning News, Wood River Times* (Hailey, Idaho) and *Daily Ardmoreite* (Ardmore, Oklahoma). Census, vital and immigration records also played a part in tracking the two men through the decades. Books sources that were helpful and, at times, even enjoyable reading are listed here:

Baten, Anderson Monroe. *The Philosophy of Life.* Garden City, NY: Garden City Pub., 1930.

Baten, Anderson Monroe. *The Philosophy of Success.* Kingsport, TN: Kingsport, 1936.

Baten, Anderson Monroe. *Why Are You Standing Still?* Kingsport, TN: Kingsport, 1934.

Biographical Directory [of] Tennessee General Assembly, 1796-1969: Hardin County [and] Lawrence County. Nashville: Tennessee State Library and Archives, 1971.

Clark, James A. and Michel T. Halbouty. *The Last Boom.* New York: Random House, 1972.

Cubage, Granville. *Oil, a Handbook for Reference: A Study of Lloyd Oil Corporation of Fort Worth, Texas.* Fort Worth: Author, 1931.

D'Easum, Dick. *Sawtooth Tales*. Caldwell, ID: Caxton, 1977.

George, Julia. *Fort Worth's Arlington Heights*. Charleston, SC: Arcadia, 2010/

The Goodspeed History of Tennessee. Nashville: Goodspeed Pub., 1886.

Gorton, George. *A "Big-ass Boy" in the Oil Fields: My Adventures, 1918-28*. New Berlin: Author, 1975.

Harter, Harry. *East Texas Oil Parade*. San Antonio: Naylor, 1934.

Joiner, Columbus Marion. *Love Letter to Heloise*. Dallas: Geyer, 1929.

Olien, Roger M., and Diana Davids Hinton. *Easy Money: Oil Promoters and Investors in the Jazz Age*. Chapel Hill: University of North Carolina, 1990.

Presley, James. *A Saga of Wealth: An Anecdotal History of the Texan Oilmen*. Austin: Texas Monthly, 1983.

Publisher's Note

Call me gullible...or maybe just human. I could probably say the same about you.

I've always believed the lore of the East Texas field discovery. I believed it mostly because I *wanted* to. I wanted to believe that Dad Joiner was an old man full of faith and pluck. I wanted to believe that he brought in that damn field with just those two things, with the help of an honest driller, some good ol' East Texas farmhands and a self-taught geologist. It's a nice story, and a popular one. And it's the one that I had in my head when I decided to add Joiner and Lloyd biographies to Harry Harter's rare history of the field. I entered this project a believer and I exit this project, as I write this, with a yearning to beat the living hell out of both men.

From time immemorial, people have tried to get their hands on other people's stuff. Those who will not earn an honest living still have to put food in their mouths somehow, most often at the expense of others. Every type of boom calls this morally skewed element of society forth from its filthy hideout and begs it to come play. I get that. I also get that not all men enter into marriage with the intention of honoring vows, and not all sire children with the intention of being fathers. There are some very unsavory characters in this world. These are mere facts of life.

I had a tough time, though, discovering bit by bit, year by year, that the two men who gave us the Black Giant—two men who have been regarded historically as heroes—were actually *anti*-heroes who lied, cheated and swindled their way into history. Dad Joiner was a low-key con artist who ditched the mother of his eight kids for his young secretary. Doc Lloyd was a loud obnoxious pig of a man who kept multiple wives simultaneously and began bamboozling for a living before he'd scarcely reached adulthood. After being forced to bring in a producing well on the Bradford lease, both men manufactured detailed biographies about what honest, hard-working gentlemen they had always been... how they'd been knocked down repeatedly but kept on keeping on. We have believed those chimerical stories for decades.

It is my sincerest wish that a better understanding of Dad and Doc will make the story of this grand field more interesting, rather than take the shine off of it. Sure, it's less of a feel-good story, but history isn't here to make us feel good. There have to be bad guys to bring out the best in the good guys. C. M. Joiner and A. D. Lloyd were pretty bad guys, indeed, but had they not stolen their way into East Texas, the historical landscape of the nation (and the world) would have been painted with a different brush. World War II and post-war prosperity, both fueled by East Texas oil, would have worn very different faces were it not for the deeds of these two men, regardless of their intentions.

Special thanks to Daniel Perez for being wise and caring enough to be edited back in a year later. Your innate ability to learn and grow without fear of failure is inspiring.

To Jim Lago who, once upon a time, had a fair amount of East Texas crude on his boots, this book is warmly dedicated. You are the anti-Dad and the anti-Doc, a straight-shooter with a keen sense of what is right and the courage to give voice to it.

-Michelle M. Haas, Managing Editor
Windy Hill

Columbus Marion Joiner as he appeared in 1889, almost 30 years of age, serving in the Tennessee House of Representatives.

BACK STORY:
THE SWEET OLD MAN

On March 27, 1947, an elderly Alabama native, then residing on Mockingbird Lane in Dallas, passed quietly into history. He had done most things during his lifetime in a quiet, unassuming manner. That demeanor made people trust him. He didn't drink or swear. He quoted the word of God and dreamed of being an author, of spreading a message of love through his poems. He played cards with poor farm widows and knew Haroldson Hunt before he was H. L. Hunt. You may know him as "Dad" Joiner. Those who bought into his myriad oil schemes across the decades knew him as C. M. Joiner, for that was the name penned to their watered down stock certificates and lease shares.

The story of Joiner's life, both before and after the discovery of the monumental East Texas field, have come down to us mostly from Joiner himself. Respected authors and historians have shared the Joiner version for decades. Oil giants of the written word, like James A. Clark and Michel T. Halbouty were there during the East Texas boom and had the advantage of having their eyes and ears in the field. Modern historians do not have that advantage. But the benefit of researching and writing on the topic 80+ years later is that vision can't be clouded by the charms of the so-called "old wildcatter."

We credit Columbus Marion Joiner and his sidekick Adelbert Durham Lloyd with discovering the Black Giant…and rightfully so. That really happened. Joiner held the leases and drilled the discovery well. But was he truly after oil? Was the Daisy Bradford No. 3 the magnificent culmination of decades of poking holes in the ground praying for a gusher?

Joiner lore tells us that the man was a true wildcatter—sinking all of his funds into the sinking of wells in unproven territory, playing hunches, taking risks, gambling everything—until he was well into his 70s. The story goes that Mr. Joiner had been wildly successful in real estate in Oklahoma, but was wiped out in the Panic of 1907. He rose from the ashes, dusted off his modest clothes, squared his arthritic shoulders and began again…this time searching for oil. Depending on which choir he was preaching to, he either discovered the Seminole and Cement fields in Oklahoma outright, or almost discovered the same but ran out of money or luck to drill just 100 feet deeper to hit pay.

Dad Joiner very much wanted to be portrayed as an everyman, a hard-working family man who finally made good in East Texas with unshakable faith and prodigious determination as the only winds to fill his threadbare sails. He had split rails, picked cotton, toiled in the fields, run cattle, read for the law and, eventually, became an oil millionaire in 1930. But Joiner lore leaves out a great deal.

So let's meet Dad.

Columbus Marion Joiner hailed from Lauderdale County, Alabama. He entered this world there on March 12, 1860 near a small village called Center Star, the youngest child of James E. Joiner and Lucia (Lucy) Rust. The family wasn't a particularly large one for the time…just Columbus and his two older sisters…since James died during the Civil War and Lucy was dead by 1870. Joiner stuck close by his oldest sister, Amanda (Mandy) Hill, eight years his senior. When she and her family relocated to East Texas in the 1870s, he followed. He took a stab at running cattle, but soon returned to Alabama where he helped out on the farm of some of his older Joiner cousins. Rooming at an adjacent farm was his second cousin, Lydia Ann Beavers. Lydia, too, had lost her father in the War Between the States. She became Mrs. Columbus M. Joiner on January 13, 1881.

Mr. and Mrs. Joiner moved up the road 40 miles to Lawrence County, Tennessee to begin their married lives. Dad Joiner opened a liquor store in Lawrenceburg and became a father for the first time in 1883, with the birth of John Lee Joiner, who later in life would become an enterprising oilman.

A few years later, in 1888, he and Lydia welcomed their first daughter, Willie May, and Joiner was elected to represent Lawrence and Wayne counties in the Tennessee legislature in 1889. He would only serve a single term… his lifelong career as a promoter had begun. He had become involved with real estate speculation deals as early as October of 1890, with his partner C. L. Norman. Their Lawrenceburg Land & Mineral Company, defunct by 1894, took out big ads in Kentucky newspapers offering shares in the company, which they deemed "the most equitable town company ever organized."

As Dad Joiner and Doc Lloyd both knew, the best way to get favorable press and the best ad space was to make nice with an editor. Barring that, the next best way was to start one's own newspaper. In 1897, Joiner founded the *Lawrenceburg Advertiser* and installed himself as editor. It was a short-lived weekly rag of 4 pages. While the family breadwinner struggled to figure out what he wanted to be when he grew up, the Joiners had added more children to their brood—James Bert, Verne Snow, Fannie (Dad's favorite), and Mattie by 1900.

Here, Joiner lore and the truth part ways, and they don't meet again for a good long while...perhaps never.

In all of the newspaper interviews that followed the 1930 discovery of the East Texas field, everyone wanted to know Joiner's back story. He gave many versions of it, but he consistently claimed that his sister had married a Choctaw Indian and that he relocated his family to Oklahoma around the turn of the 20th century to help her manage her land allotment under the Curtis Act. The other members of the tribe were so impressed with his legal acumen, so the story goes, he helped them all out and began a robust career in Oklahoma real estate. There are two problems with this particular bit of Joiner's story: he had no such sister and he was probably not a lawyer.

Joiner had two sisters—Amanda (Mandy) Hill and Margaret (Mamie) Simms. Both married gentlemen who were decidedly un-Choctaw. Mr. Hill's folks were from the Carolinas and Mr. Simms' were from Tennessee. Likewise Joiner's only sister-in-law. That Joiner struck out in promoting Tennessee real estate and wanted to start over in Oklahoma would not have made juicy interview fodder. That story didn't paint him as the generous soul he wanted to be seen as, and he didn't have the natural panache to be believable as a high fallutin' real estate mogul. The real story of how he came to settle in Oklahoma was a common one for the time, and boring as hell, so he revised it.

Joiner became a lawyer or at least began studying law at some point in the 1880s, depending on which history you're reading. *The Handbook of Texas* (online edition) and *The Last Boom* have him practicing law as early as 1883. In 1887, however, when Joiner provided his own biography for the "mug book" portion of Goodspeed's *History of Tennessee*, he did not mention having pursued any legal career at all. To read for the law, serve an apprenticeship and be accepted to the bar took years out of a man's life. There is no evidence to support the story that Joiner took the time out to become a lawyer.

C. M. and his middle son, Bert, are found in the 1900 census, in Ardmore, Oklahoma, *not* rooming with family that supposedly was there already or with the Choctaws that Joiner claimed to have been so intimate with. Ardmore, after all, was Chickasaw country. Rather, they were boarding with an upholsterer from Arkansas. On this census alone did Dad Joiner claim to be a lawyer by profession. Unless he was running for local political office or giving his life story after the 1930 oil discovery, he was never a lawyer again. Had he truly been, he might have attained in Oklahoma what he spent the next four decades in search of. He was, after all, in the home of the land rush, the Boomers, the Sooners, the Eighty-niners—a place destined for statehood and poised for growth. There was money on the table for the legal man. Yet Dad Joiner never hung his shingle out for that trade.

While father and son scouted business prospects and cheap land in Ardmore in 1900, Lydia and the remainder of the children were in a rental house in Tennessee waiting for Dad to send for them. By the end of 1901, she and the brood had arrived. Soon Joiner's sisters, Mandy and Mamie, with their families, assembled there as well. Columbus Joiner had begun again in Indian Territory.

His first move as an Oklahoma real estate promoter was to launch a newspaper. In February 1901, the *Ardmore Appeal* came into being and sixty days later, it was no more. Joiner had chosen a local man named Peirce as his partner in the real estate business. They had barely begun when, in August 1901, Joiner ducked out for two months and headed to a health spa for whatever rheumatic ailment afflicted him. Upon returning, he became a notary public and by late 1901, Joiner and Peirce were set up in bank offices. They took out ads and began promoting themselves as big real estate men. They'd sell your land or help you buy a piece; they'd sell you fine grazing land in Mexico; and, of course, they'd sell you land that they were buying on time from somebody else.

In one of their early transactions, they sold a parcel of land for one James Pulliam for $400 (roughly $11,000 today). Mr. Pulliam received only piecemeal payments from the proceeds and partial payment would not do. He took Joiner and Peirce to court in July 1902 and they were ordered to pay. But the real estate boys had other plans.

They were moving into bigger, fancier offices in the brand new Noble Brothers Building before the wallpaper paste was even dry.

On February 24, 1903, Joiner - now appending the title of "judge" to his name - announced that he would run for the office of Police Judge of Ardmore. His credentials included being "a lawyer by profession and a former member of the newspaper fraternity." He was running against the incumbent, John Galt, and lost miserably in the April election. Lydia and the younger children (there were a total of eight Joiner offspring now) headed back to Lawrenceburg, Tennessee for about 6 months to stay with relatives. Joiner found himself a new partner in the real estate racket, J. W. Hoffman.

With Hoffman and two local Lester brothers, C. M. Joiner formed the Commonwealth Land & Trust Company in 1904, with a capital stock of $100,000, to engage in the banking and loan business. They claimed that half of that had been paid in and quite a dubious claim that was. This same group threw their hats in the ring in early 1905, when the Ardmore City Council was soliciting bids to have an electric line built to power up the growing town. The Council, after much hemming and hawing, awarded the contract to Joiner, et al. Building was to begin by September and was to be completed in one year. Although the Lesters and Mr. Hoffman stuck by the project, Joiner pulled out and began a new real estate partnership, this time with a man named Keller. This partnership, too, would be very short-lived.

Foreseeing statehood and aiming to get a foothold in local leadership, Dad Joiner announced a run for an alderman spot on March 8, 1906. But without having been active and trusted in the Ardmore community of about 5,500 citizens (and without campaigning), he was lagging and dropped out of the race 8 days later. Meanwhile, back at the several rental homes that the Joiner clan moved between from time to time, Lydia was looking after their 8 children. If the social columns of the time were any indication, Lydia Joiner didn't get out much.

Columbus was cooking up a new plan to get rich. He had been buying up lots on time, about twenty miles from Ardmore and planned to promote his own town, which naturally would be called Joiner City. Still trying to hook in to respected Ardmore society and have his name associated with things lofty and important, he traveled to Dallas at his own expense to attend the big Democratic convention there and report back to the local paper. He joined the local Elks Lodge to do some networking. He judged local horse shows and participated in a spelling bee or two at the local school (teachers vs businessmen.) He tried to be the man-about-town, but with poor cash flow and no big successes as a businessman, he had little to inspire the confidence of his neighbors to convince them to leave Ardmore and move out to his development. But Joiner had land that he could offer on payment plans, so he was popular with certain segments of Ardmore society. He sold a few lots and Joiner City began to gain momentum.

An event that shaped the lives of the Joiner clan around this time—a personal one that gets left out of Joiner lore entirely—occurred just before Oklahoma statehood in 1907. Bert Joiner, Dad's middle son, was accused of shooting his girlfriend in the head after a drunken argument. Former partner Hoffman and new real estate partner, Wolverton, along with Joiner and a few others, posted Bert's bond. The trial wasn't a pretty one and revealed that Bert was keeping company with "joint" owners and drinking excessively. (Oklahoma was dry at the time.) He was acquitted of all charges in a few months, the jury concluding that the girl shot herself because Bert was quitting her. Once exonerated, Bert helped Dad sell lots in the Joiner City land development. Bert's worst troubles were yet to come.

Although Joiner lore tells us that he had amassed a real estate fortune by this point, then lost it all in the Panic of 1907, that seems untrue all the way around. The remainder of the decade was business as usual. Property he owned outright continued to get sold in sheriff's auctions because he failed to pay property taxes. People from whom he was buying land on time or from whom he had borrowed money sued him. But the tax sales and lawsuits had been routine happenings since the Joiners came to Oklahoma. Dad ran for mayor of Ardmore in 1909 but lost in quite a dramatic fashion. He filed the plat of Joiner City that summer and made the rounds to neighboring towns talking it up. Then, poor Dad took a month off to travel to Chicago and Hot Springs to visit health spas. He made a trip out to Oklahoma City, bought himself a new car and drove her home. The Panic of 1907 didn't wipe him out—spending money that he didn't have kept him wiped out until the day he died.

While the delinquent taxes and lawsuits piled up, Joiner City was beginning to thrive. It had a new cotton gin and its own voting precinct by 1910. It even boasted a 20-piece band, called the Springer Band, that participated in Carter County parades and other entertainments. But Mr. Joiner rarely visited the burg that bore his name. His real estate partner, Wolverton, took over most of the Joiner City development because, once again, Dad was distracted by potentially greener pastures. He had discovered the syndicate stock scam and oil was nearby. The rutted and winding road to East Texas was now being paved for C. M. Joiner.

The Glenn Pool, brought in in 1905, ushered in a crude-crazy era in the Oklahoma and Indian Territories. Everybody wanted a gusher. Some prominent Ardmore businessmen raised some local cash to begin poking holes in the ground nearby. Joiner caught wind of this and chartered a new corporation in late 1909 with his real estate partner, J. J. Wolverton. It was called the Joiner City Oil & Gas Company, capital stock $250,000. The company sat idle until the men from Ardmore began actively drilling near present-day Healdton in 1912.

This lit a fire under Dad Joiner. He began running in Oklahoma City circles, spending less and less time in Ardmore. In 1913, after the discovery well of the Healdton Field came in, he created a second corporation, Few Acres Oil & Gas, capitalized at $35,000 and organized in Oklahoma City. As luck would have it, Healdton was within spitting distance of Joiner City and C. M. wanted to cash in on the discovery…not by sinking wells, but by snatching up a cheap lease or two, then selling shares of those leases and stock in his companies. His OKC partners peddled stock in Few Acres, while he worked the Ardmore territory for Joiner City O&G and sent another man on to Shawnee and other points to peddle shares.

The Healdton Field also renewed Joiner's interest in the little village that bore his name. He advertised free land in Joiner City to anyone who would build on any unsold lot there. Why? Magnolia was building a pipeline into Healdton. The Oklahoma, New Mexico & Pacific Railway was building a rail line to haul the supplies and pipe. Supplies had been terminating at Ardmore's Santa Fe station and then moved by carts to the field. A new rail terminus closer to the action would quicken its development. Joiner City needed to give the appearance of a fully-formed town in order to lure the railroad. Announcements were made about the big new office building Joiner City Oil & Gas was erecting. All oil field traffic was to flow through Joiner City as soon as the rail line was completed, Dad announced to his news-hungry friends at the Ardmore paper.

Columbus had some stiff competition, though. John Ringling was using some of his hard-earned circus money to establish a terminus for the railway himself. In addition to the cash, Mr. Ringling had another advantage in landing the terminus—*he* was the one building the railroad.

Joiner did not let these little details deter him, though. He filed a complaint with the Oklahoma Corporate Commission stating that his town had already been platted, whereas Ringling's town was not. The Commission found that Joiner City was, in fact, the most convenient location to serve as a station for freight purposes. On January 14, 1914, they ruled that Ringling's railway must put in a spur or regular switch to run the pipe shipping through Joiner City to accommodate the "emergency matter" of "taking care of the necessities of the oil industry." The decision, however, carried the caveat that if the field should expand westward, Ringling's proposed town would be the more logical terminus. Ringling's railroad had less than month to comply.

So the circus magnate brought the matter before the courts. The field was expanding and he needed to buy time. By the summer of 1914, the case had landed in the lap of the Oklahoma Supreme Court. While the wheels of justice chugged along, Ringling filed the plat for the town that bore his name and the Healdton field expanded westward. The Corporate Commission, of course, ruled that the new town of Ringling was the official terminus for the railroad and Joiner City was out.

In the months between rulings on the railroad issue, Joiner and his associates were busy trying to sell shares of their two companies. They had been treading lightly at this early stage, relying on personal visits to prospective local investors eager to get in on the oil action. Dad's youngest son, Verne, was now buying and selling leases on his own account. His middle son, Bert, was in the papers again – not for murder this time, but for running alcohol in dry Oklahoma. The failure of the Joiner City venture, coupled with moderate success in stock sales pushed Columbus M. Joiner out of Ardmore and into Oklahoma City and the life of an oil promoter for good.

Joiner's son-in-law, W. J. Lane, husband of daughter Willie, had built a reputation in Ardmore as one of its most straight-dealing and enterprising citizens. He built a thriving grocery business from scratch in his late teens, and by 1915 was very comfortable. Dad Joiner leveraged Lane's reputation by installing him as president of his latest promotion scheme: Oklahoma Star Oil Company. As secretary and treasurer of the company, Joiner was responsible for snatching up the money and issuing the stock. Joiner announced publication of a news sheet for the Healdton field that was, in truth, just a promotional tip sheet to promote OSOC. The Joiner City Journal did not take off, but it was a sales tactic that Joiner would employ again down the road.

By the summer of 1915, Columbus M. Joiner was calling Oklahoma City home. Lydia and daughter Fannie held down the fort back in Ardmore. His Oklahoma Star Oil Company was advertising stock shares more heavily in the newspapers of far-flung places like Pennsylvania and Wisconsin, as well as feeding stories to the Ardmore newspaper about the planned activities of the company. There were, after all, a large number of investors in Ardmore who needed to believe that the company was on the level. Occasionally, a news story would pop up about a well being drilled or plugged. Rarely, production was reported, and these reports topped out at 5 to 15 barrels a day and were short-lived...just enough to show investors that oil was flowing, but not enough to justify paying the dividends that were promised.

Just before the Christmas of 1915, Joiner brought a carload of Ardmore investors with him down to Shackelford County, Texas to watch him buy up some land there. It is probably closer to the truth to say that he secured a lease on the land, if even that. It was far enough from Breckenridge and Ranger to likely not have been productive even if he had drilled. The important thing, of course, was that the people of Ardmore heard that Joiner was out and about tending to important oil business.

Enter Joseph Idelbert Durham, a.k.a. Adelbert Durham "Doc" Lloyd, a.k.a. Adolphus Delbert Lloyd...

Joseph Idelbert Durham, a.k.a. Adelbert Durham "Doc" Lloyd, circa 1919, nearing 60 years of age.

Back Story:
The Good Doctor

Just as there is Joiner lore, there too is a turbulent and confusing world of Durham/Lloyd lore. Known to us today as the geologist (or wannabe geologist, in more accurate histories) who located the East Texas discovery well for Dad Joiner, Lloyd brings a flash of color to the East Texas tale. The annals of Texas history are filled with counterfeit colonels and other men of questionable credentials. Durham/Lloyd's place is secure among their ranks in perpetuity. He belched out lies about his past in such a loud and confident manner, it seems there was hardly an opportunity for the listener to *want* to disbelieve him. While Dad Joiner tried to look like the honest, modest common man to win over the people, Durham/Lloyd wanted the world to believe that he had been to the moon and back, found oil on that celestial body…and had done it with a beauty on each arm, feeding him delicious fried chicken all the while.

After the East Texas discovery, Lloyd's story as he told it was riddled with falsehoods and tall tales. Surprisingly, the least of those was his self-applied title of "doctor," which had many meanings over the course of his lifetime: dentist, physician, pharmacist, chemist and geologist. He liked to brag to the East Texas folks that he'd been married six times and that may be the only detail of his life as given by him that flirted with the truth.

He claimed he was born in lawless and feud-ridden Breathitt County, Kentucky, where he fought his first duel at the age of nine and where his father was likewise killed when he was just twelve. In another version, he is defending white settlers against Indians at Yellowstone at age twelve. He claimed that his Kentucky kinfolks sent him off to Cincinnati afterward to save him from assuming the bloody responsibilities of the eldest son of a feuding family. In pre-East Texas versions of his life, he had served in the Spanish American War as a special scout who damn near fought the entire ten-week conflict single-handedly. The version he used during his time spent in Idaho was that he raised the first regiment of black volunteers to serve in the Span-Am War and was promoted to the rank of lieutenant-colonel. (After WWI, he dropped the war hero line from his repertoire.) Further tales told to the folks of East Texas were of gold prospecting in Alaska, Idaho and Mexico and of being a soldier of fortune in the latter place. The truth about Joseph Durham's path to East Texas is far more entertaining to modern ears than even the versions he told.

Joseph Idelbert Durham was the firstborn child of Alonzo Durham and Mary Markley in the village of Cherry Grove just outside of Cincinnati. He arrived on May 14, 1870, a big boy with blue eyes and dark hair. Although many histories tell us that he was older than Dad Joiner, he was actually a decade younger. He had 3 younger sisters and the family engaged in typical agricultural pursuits. By the time he reached age 15, his father had passed away and Joseph was working as a clerk in a pharmacy in Cincinnati, where he lived with his widowed mother and little sisters. After clerking in the pharmacy a few years, he began a side business—that of practicing medicine. He obtained a mail order medical diploma and began calling himself a doctor. In the spring of 1890 he joined the 14th Ohio National Guard. His enlistment paperwork indicates that he was of average height—a hair under 5' 8"—and not the towering giant he is made out to be in modern histories of the East Texas field. He claimed that he had been formally trained as a surgeon, so he was assigned to the hospital corps. He was honorably discharged just as 1892 rolled around. During the time of his Guard service, he obtained several more mail order medical licenses and actively practiced quack medicine in Cincinnati.

Now Joseph Durham begins to take the shape of the larger-than-life character of East Texas fame. In 1892, he headed out west. He was not digging for gold in the conventional sense…he was rifling through the pockets of the gullible. He became a snake oil salesman of the highest order as a distributor for the Kickapoo Indian Medicine Company. To proffer their Sagwa remedies, as well as electric belts and other quack staples of the day, he set up a traveling medicine show, dubbed Dr. Alonzo Durham's Great Medicine Show. One wonders what his deceased father might have thought of such an homage. Durham had in his employ 2 Indians and 2 Negroes who provided the entertainment portion of the show. They'd do a song and dance, then he'd hawk a tonic or unguent. Another ditty, then another product. By the end of the evening, if anyone remained, he would sell a pamphlet containing recipes for his own concoctions. He wended his way through Utah, then up into Bear Lake County, Idaho, then on to Shoshone where something resembling justice awaited.

During a two-week stop in Shoshone, Idaho in September of 1894, he had cleared about $1,000 in sales. But when the Idaho Medical Society asked to see his credentials, he could only present the mail order diploma from the American Eclectic Medical College of Cincinnati. So Durham was arrested in Shoshone, where he made his $150 bail and awaited trial. In the meantime, he tried to find an angle that he could work to his advantage. The Censor for the Medical Society and the man who reported him, Dr. N. J. Brown, practiced medicine in a town called Hailey on the edge of the rugged Sawtooth Mountains. During the trial, Durham's attorney made the plea that Dr. Brown did not have his diploma on file with the proper county authorities to practice medicine in Hailey. It had only been recorded in an adjacent county, which left the Censor himself open to arrest for practicing without credentials on file.

The local paper, *The Wood River Times*, was abuzz with the news of the medical diploma scandal and had interviewed "Dr." Durham immediately following his arrest. Seeing that Hailey was about to be sans a doctor if Brown was arrested, Durham seized upon the opportunity to squeeze into the open slot. The best office space in town, of course, was where the people received and reported their news, so Durham rented the rooms on the second floor of the *Times* building and made fast friends with the editor. The free publicity (and the barrage of lies) began at once. Reporting on his arrival, in the October 20, 1894 edition of the *Times*, the faux physician's resume was given in glowing terms:

Dr. Durham is a graduate of the American Medical College of Cincinnati and is also a graduate in pharmacy. In 1892-3, he held the chair of Pharmacy, Chemistry and Toxicology in the Medical University of Ohio. He is, therefore, not only a physician but a pharmacist and full-fledged professor, as well. Dr. Durham has extensive and varied experience in travel and practice in all parts of the United States.

The same paper that had reported that Durham's diploma came from a school that wasn't a school at all now embraced him as a bona fide physician and, better yet, a paying tenant. The new local doc began advertising $5 catarrh cures, which included all medicines, provided the patient clipped and presented the coupon in the ad. This assured his editor friend the sale of papers as well as it assured him patients. Dr. Brown was never arrested but a rivalry remained. Durham was released after the trial after filing a second bogus medical diploma with the proper authorities.

To drum up publicity for Durham and to fill space in the paper, sometimes the *Times* would manufacture sensational medical news. In early February 1895, a story ran about a "famous" cowboy and renowned snake handler, Rattlesnake Jack, who was kicked by his horse and wounded in the lower abdomen and groin. He could well have gotten treatment from doctors much closer to his camp, he said, but "the range riders [had] no confidence in them." Instead, he chose to travel 180 miles to Dr. Durham because his charges were reasonable and the good doctor would allow him to make payments. Several articles ran about Jack's snake handling prowess and his encyclopedic knowledge of rattlesnakes, along with tidbits about how Dr. Durham was treating him so well as to return him to the saddle as quickly as possible. Rattlesnake Jack, the rough-riding cowboy, was actually John Kline, a mailman. None of the lengthy articles about him disclosed the actual nature of his injuries or how Dr. Durham went about treating them.

In March of the same year, an announcement was made that Durham had received $500 in new medical and dental equipment and would commence practicing dentistry in addition to quack medicine. Said the *Times*, "Dr. Durham has persistently refused to practice dentistry…But he has finally yielded to the solicitations of his patrons and will devote to dentistry what few moments he can spare from his medical practice." The following month, he was advertising in the Salt Lake City newspapers seeking an experienced, young female nurse to add to his practice. Graduates of legitimate nursing schools were preferred, but age was more important to Dr. Durham than the gal's references.

The rivalry between Dr. Brown and Dr. Durham was still smoldering in April of 1895. The citizens of Hailey were divided between the reliable, legitimate Dr. Brown and the sensational, boisterous Dr. Durham. But Durham had the editor of the *Times* in his corner and when someone attempted to make a mockery out of Durham's quack practice, his portliness and his attire, there was an outcry from the newspaper. Dated April 17, 1895, bearing the headline "An Outrage," the story served as a warning to those who might mess with the editor's pal in the future:

Last night, some vile scrub put up a dummy of a man on a pile of lumber that was hauled yesterday upon the lot where Dr. Durham is having his stable erected. The dummy was intended to impersonate a fat man. It was made of white duck, stuffed and topped with a gray campaign hat. The intention of the thing was to reflect upon Dr. Durham. But it failed at such effect, as it was bound to.

Dr. Durham came here only seven months ago. But he has already secured such a practice as most physicians never get in a lifetime. Every patient he has treated should be his friend and probably is. He pays his bills promptly, is so busy that he cannot meddle with others' affairs if he would, and is certainly the peer of any in this community in attainments and promise.

This talented young man—for, though he has had an enormous amount of experience in every branch of medicine and surgery, Dr. Durham is not yet 30 years old – need ask no odds of anybody; nor can he be injured by the petty souls of the few pompous, dishonest semi-idiots who imagine that they own this community. He is here, he actually cures his patients, he is very reasonable in his charges, and he is going to continue to prosper in spite of all the dirty work that has been or will be done him. He had, therefore, better not be interfered with.

All in all, it appears as though the future Doc Lloyd had a pretty cushy life during his stay in Idaho. He was surrounded by some of the richest beauty on God's green earth. He had free unlimited advertising by the local newspaper. His office was open for an hour each morning from 9 to 10, two more hours in the afternoon and then generously from 7 to 11 p.m. He had an assistant, a driver and two teams of horses. His office and apartment were festooned with lace curtains and velvet carpets. He had all of the necessary equipment to make his practice look legitimate and to compound his own so-called medicines. The editor of the *Times* even ran above the paper's own masthead for a month an announcement that "Dr. Durham Compounds His Prescriptions Himself." Not bad for a guy without an ounce of medical training. Something is missing, though. What's a good story without a love interest?

On July 8, 1895, a story ran that shocked Hailey, Idaho. The young doctor had married on the night of July 7, quietly in the home of the editor of the *Times*. The bride – his first, as far as we know—was Miss Jeanette "Jennie" Klotz of San Francisco by way of Yankton, South Dakota. She had arrived on Independence Day with the intention of marrying Durham in a month's time. But upon her arrival, the two decided an immediate wedding was in order. The paper jazzed up the bride, suggesting that she was a native of Europe and made $100 a night playing violin in the orchestra of her brother, Professor Louis Klotz. How exotic! While it is true that Jennie came from Bohemian parentage, she herself was born in Chicago. Her brother, though he took euphonium lessons and made a living gigging as a baritone sax player, was no more a professor than was Dr. Durham.

There were but a few more news items in the *Times* about the editor's pet doctor before what he described as a "public calamity" occurred. Doc Durham jumped town. He sold all of his horses, office furniture, land notes, accounts receivable...everything but a few instruments and his wardrobe...to the *Times* editor and high-tailed it out of Idaho. The story ran on August 26, 1895. Durham claimed that he just couldn't get enough rest with his practice as busy as it was. And besides, even though he was making about $1,000 a month, he was only collecting about half of that amount. He said he hadn't decided where he was going, only that he needed to go. He figured a man with his training could earn $20 a day elsewhere with dentistry alone. With that, he was off. The editor of the *Times* attempted to sell off the doctor's personal affects without much luck. By September, everything he'd left behind except for the livestock and land, were put on the auction block. Rumors, likely seeded by Durham himself to add flavor to his hasty exit, flew that someone, perhaps another sweetheart, had taken a shot at him through his bedroom window as he slept.

In an interview with the *Times* about a decade later, Durham claimed that after only 16 months of wedded bliss, his first wife lost her appetite and died. But Jennie was probably sick when they left Hailey, and that likely had a good deal to do with why they left. It wouldn't do for the doctor's wife to be ill and for him to demonstrate that he had no idea how to treat a sick person. She was dead by the end of 1895. He buried her in South Dakota, then made tracks for home.

Back in Cincy, the former "Doctor" Durham began working in a pharmacy once again until he could put a deal together. He found two young men who would do the song and dance portion of the program while he did the snake oil selling. He took the show on the road, making his way up near Lake Erie to Adrian, Michigan. In 1902, he met the Second Mrs. Durham, a widow named Anna Rose Meeks. She had two half-grown children, Lillian and George, from a previous marriage. It is presumed that the children are from a previous marriage because they were born in Georgia and Alabama, respectively. Air travel might have made it possible for Durham to have sired Anna's older children and return to Cincinnati to practice medicine, but planes were not available to our fertile pseudo-doctor in the 1890s when the children were conceived.

The couple settled in Adrian, Michigan temporarily and Joseph Durham plied his medical and dental trades again. A son, Ricardo, was born to them there in the spring of 1903. In 1905, Durham was "curing" patients with radio waves. He authored a pamphlet called "Durham's Radio Therapeutics" during his time in Michigan, hoping to get rich with the radio therapy trend. When that didn't work, he moved the family on down the road south to St. Louis, where he deposited them and then kept right on moving. When he reached his destination, he carved out a new career and a new name for himself.

The destination? Fort Worth, Texas.
The career? Arlington Heights subdivision promoter.
The name? Adelbert Durham "A. D." Lloyd.

And just like that, a new Texas character was born.

What possessed Joseph Durham to become A. D. Lloyd in 1906 and head to Texas is not known. In stories he told after the 1930 East Texas discovery, he left out his real estate career entirely. Arlington Heights had waxed and waned since the 1880s. Perhaps he had seen a newspaper ad for the subdivision. Maybe he knew an investor. He, by chance, could have heard about it at the 1893 Chicago Columbian Exposition, where promotional picture books had been distributed. Regardless of why he came, he dove into his new life head-first. Not only was he buying up a few lots on time, he also took advantage of a new building craze and started the Arlington Heights Concrete Block Building Company. It wasn't long, however, before chasing real estate deals eclipsed the building business. Though

Lloyd later told an Idaho newspaper that he owned all of Arlington Heights, he was merely a sales agent. The real money invested in the subdivision came from his boss, J. Stanley Handford.

As the financial crisis of 1907 drove lot prices in Arlington Heights from $300 to about $50, Lloyd's commissions slowed. In the first months of 1908, he paid a visit to Anna and the kids back in St. Louis. In November of that year, another son was born and was named Douglas Adelbert Durham. His daddy was already back in Texas by the time of his birth. It was during his Arlington Heights Realty Company days that Lloyd met George E. Montgomery, a gambling man of odious repute who would enter the oil game with him later in Oklahoma.

Out of the East Texas field's cast of characters, the most successful and enduring name is probably that of H. L. Hunt, who built an empire from his successes in the field. Compared with what Hunt built and sustained, Joiner and Lloyd are but second-tier swindlers. That said, A. D. Lloyd did give Texans something of a lasting nature—something vile or something valuable, depending upon which side of the fence you're on—the homeowner's association. As a promotional gimmick, Handford and Lloyd sold lots tucked safely behind stone entryways, where buyers agreed to build homes of at least a $10,000 value and commercial interests were strictly disallowed. The association was called Beautiful Arlington Heights and the idea of such restrictions in suburbs caught on and has stayed on.

While Lloyd was away in Texas, Anna's two teenage children, George and Lillian, were sent to good schools. Lillian would be schooled in the arts in Washington State. She would become a painter and a poet who wrote one novel and a few songs about Texas with her brother George. Lloyd's first son with Anna died at age 50 in California after a lifetime of alcoholism. Their second son, Doug, figures into our story later. It does not appear that Joseph I. Durham ever legally changed his name to Adelbert D. Lloyd. All of his children were legally born Durhams even after he assumed his new identity. Explaining the name change to his wife and kids in St. Louis must have made for interesting conversation around the family table when daddy came to visit. Lillian chose the Durham name. George did the opposite and called himself "Tex" Lloyd. Ricardo remained a Durham and Doug was a Durham until about age 10 when he was enrolled in school as Douglas Lloyd.

A. D. Lloyd hung around Arlington Heights until the end of 1912. Stepdaughter Lillian was to attend art school in Washington and Lloyd jumped at the chance to head out west again. After delivering Lillian to school, he revisited the town where he had honed his skills in quackery. In April of 1913, he stopped a week in the sunny mountain town of Hailey, Idaho and received a hero's welcome from his old friend at the *Times*. Durham-turned-Lloyd was, of course, interviewed by the paper. His flair for the braggadocio and bullshit that he would be known for in East Texas shines through brilliantly as he recounts his adventures since leaving Hailey in 1895:

Joseph I. Durham, M.D. arrived by the train yesterday to visit with a friend overnight. So many of his former patrons and friends have urged him to stay, however, that he will probably prolong his visit for a day and will not leave until tomorrow, when he will resume his trip to Fort Worth, whence he departed six weeks ago for a two month visit through the northwest and the Pacific Coast, in order to have a restful change of scene and to brighten his fagged ideas.

Met at the depot by the editor of The Times, *whom he specially came to see, Dr. Durham was hardly seated in the newspaper office when former patrons called on him...*

Henry Watson, Carl Hartung, Henry Ward, over a dozen other men and several women all walked up to Dr. Durham on the street and thanked him for successful treatment of them. Some even asked if he could not stay a while, to treat them or their friends; and several invited him to dinner, etc. All these advances Dr. Durham was reluctantly compelled to decline for lack of time and because he has written only one prescription in 16 years and that was for a friend suffering from inflammatory rheumatism while on a visit to him in Fort Worth.

While practicing here in 1893-95, Dr. Durham was married to Miss Jeannette Klotz of Yankton SD...Miss Klotz was most winsome and petite and an accomplished musician having for years previously played on the violin and the piano in the stringed orchestra of her brother that supplied the wealthier San Franciscans with dance music, and for which she was paid $100 per night, her brother receiving double that amount.

Her union with Dr. Durham followed an estrangement that resulted in the half of the continent intervening between the young lovers, the doctor being then a mere lad of 22 and his bride a lassie of 20.

After a year of wedded bliss, Mrs. Durham lost her appetite but did not seem otherwise ill, and the fond husband consulted several of the most eminent specialists in eastern cities but in vain. Death claimed her 16 months to the day after her wedding.

Dr. Durham took her body to Yankton for burial and then determined to quit practicing medicine. Up to that time, so pronounced had been his success that he believed himself almost infallible. This illusion rudely dispelled, he was unwilling to continue in practice and he bought first one then another drugstore in Cincinnati.

At the outbreak of the Spanish War, he organized the First Ohio Colored Volunteers in Cincinnati, was commissioned its major, promoted in the field its lieutenant colonel and he served in Porto Rico until ordered home for honorable discharge.

After vainly endeavoring to become attached to any line of business, he was induced to buy the Arlington addition to Fort Worth. He had out $17,000 but paid $15,000 down, got several years' time on the balance, a year later interested a friend who invested $25,000 for the scheme, bought him out for $100,000 a year later, then added to his holdings until he had three additions to Fort Worth and his furthermost lots were six miles from the business center of that city. He then had a stretch of city lots 2 1/2 x 4 miles in area and constituting the entire west and extreme fashionable side of the city.

Soon after he bought his partner out, times hardened, money became scarcer, and when the panic of 1907 came on, Dr.—now Mr. Durham—found himself involved in liabilities totaling $520,000. Fortunately he had accounts at 17 banks and trust companies, and by assigning his contracts of sale on monthly payments and all his property, *his financial bankers carried him at 6 and 8 per cent per annum until the 15th of last September, when he finally paid up in full by taking up the last note.*

He then looked over his property to see what it was worth. He had some land in Georgia; also 12,426 acres bought from the Ballantine estate in the Texas Panhandle; 3,900 acres in El Paso county bought from the state on 40 years' time at 3 per cent interest; 9,022 acres of cotton land in six tracts in Cherokee County; 2,775 acres in San Patricio County in the Bermuda onion belt; and much other property, the most of which he had not even seen, having acquired it in some way in course of eight or ten years' dealings, during which he had built an electric road through his Fort Worth property, also put in water works, electric lights, sewers, miles of cement walks, a telephone exchange that now has 800 subscribers, dug an artificial lake of 36 acres, built himself a $50,000 residence fronting it, and donated the property for the Arlington Heights Female College that is sustained by the Sunday League of America as a high grade finishing school for wealthy young ladies.

He also donated the property for the Baptist Industrial College, and is interested in raising $250,000 for the construction of the Southland University of the Church of Christ (Campbellite)—all in his addition; 40,000 acres of an 18-foot vein of bituminous coal in Palo Pinto and adjoining counties and much other "stuff."

Concluding that he was not in imminent danger of the poorhouse, he felt that he could afford a brief vacation. But he must now get back. So he will leave tomorrow.

Under the dateline of April 13, 1913, the issue of Durham's name change was addressed after he had skipped town:

AD Lloyd of Fort Worth, Texas, left today for home, after a week's visit here. He intends to stop for a couple of days in Hillsdale, WY to visit two of his cousins, one of whom is a minister of the gospel and the other the principal of the public schools. He will also stop in Denver for a week and he expects to be home by the 23rd instant.

Mr. Lloyd is known here as Dr. Joseph I. Durham. But when he quit the practice of medicine, 16 years ago, he sought to drop his past behind him forever. He had not been guilty of any crime, but he had conducted a show three years for the Kickapoo Medicine Company and had a similar show of his own. That show included a 15 instrument band, about 40 performers and a total enrolled of 110 persons. It was as big as some circuses.

But he resolved to settle down in the real estate business. So, obtaining an order of the court, he changed his name and made a fresh start.

Mr. Lloyd intends to return this fall with a carload of good fellows, to hunt bear and other big game in our mountains.

At this point, it's safe to wonder if H. L. Hunt was taking lessons from observing Doc Lloyd's amorous stunts. Lloyd headed back to Texas from Idaho. He had already met his third wife, Louisa May Chandler, a native of Iowa who was raised in Illinois. He set May, as she preferred to be called, up in a boarding house in Houston. He then rushed back to St. Louis and brought Anna and the kids to Fort Worth to his home in Arlington Heights. Now he had two wives in two cities. By 1914, May had given him a son, born Joseph Durham (later Joseph Durham Lloyd). Lloyd continued to dabble in real estate and worked as a travel agent. But with wives and kids in the orbits of two oil cities and the decade of the 20s rapidly approaching, it isn't difficult to surmise what his next move will be.

SWINDLERS UNITE!

As oil and gas concerns began to look increasingly toward geology as a means of locating oil, A. D. Lloyd had begun to test the waters of oil promotion. The era of shady oil promoters and stock sales was in full swing and there weren't many legal checks in place yet to throttle the con men and swindlers. Nonetheless, A. D. Lloyd chose a different route, perhaps because it carried less risk…or more likely because it allowed him to affix fake but prestigious titles to his name once again. Perhaps he missed being called "doctor." So he ordained himself a geologist. He sometimes claimed to be a chemist as well if it was convenient or necessary. Not only did he become an instant geologist, he billed himself as the best geologist in the United States and the official geologist and engineer for the Mexican government. He consulted with wildcat operators and was paid for his reports in stock certificates or the cold, hard cash needed to support his two families.

After a stop in Houston to impregnate May with their daughter, Dannette, Lloyd made visits to Oklahoma in late 1915 and talked himself up as a scientist. By 1916, he had written his first faux geological survey for the Signal Hill Oil & Gas Company of Oklahoma. They needed something flashy to sell stock and Lloyd did not disappoint. Naturally, his report on their location was favorable…they all were. He poetically reasoned that since oil on the surface of water displayed a range of colors, so too would colorful rocks on the surface of the earth indicate oil was beneath them.

In June of 1916, Lloyd's domestic arrangements were altered yet again. His Fort Worth wife, Anna, died of tuberculosis in June of that year. He placed Doug, then 7 years old, in the Bryant School for Boys in Fort Worth. It is unknown where the older boy, Ricardo, landed. Houston wife, May, and their two small children, remained in Houston, where Lloyd spent the portion of his time not spent in Oklahoma.

Although the exact moment in time when the forces of Joiner and Lloyd were united cannot be pinpointed, it certainly occurred during Joiner's Oklahoma Star Oil era. They likely came together when Oklahoma Star was drilling a well on the Kunsmueller tract in what would become the Cement field in late 1916. The earliest referenced connection between the faux wildcatter and his pseudo-geologist tells us that the two were operating together by January of 1917.

Storm clouds were gathering over Oklahoma Star as 1916 drew to a close. But Dad Joiner's shady management ultimately provided Doc Lloyd with a pretty sweet long-term job with some interesting benefits. George Gorton, a successful machine shop owner from Wisconsin who was heavily invested in Oklahoma Star, came down to personally visit his "investment." He found that the company had no intention of producing oil at all, and was paying so-called dividends to investors occasionally out of stock sales, not oil production. Mr. Gorton could well have asked that the company be put into receivership and scrutinized, and Dad Joiner knew how such proceedings could hamper his future stock schemes. Instead, when Mr. Gorton suggested cleaning OSOC's management house, Joiner bowed out.

Gorton tossed Dad Joiner out on his ear and took over. But he kept the talented and fearless "Dr. Adolphus Delbert" Lloyd on the payroll. Lloyd and his old companion from Arlington Heights, George Montgomery, had surveyed land for Oklahoma Star in Cement and were adamant that Mr. Gorton continue to drill. Clever Lloyd managed to worm his way into a steady paycheck by encouraging George Gorton's drilling activity for the next decade. With Gorton's financial backing and actual drilling, Lloyd was given an air of legitimacy and was free to continue making money writing bogus geological reports for other companies.

Gorton continued to use the Oklahoma Star Oil Company name for a few more months and when the Cement pool came in, it was under Gorton's management, not Joiner's. Gorton was hailed as a hero in Caddo County and Doc Lloyd was at his side.

Meanwhile, Dad Joiner had begun another stock-selling company, White Rock Oil Company. WROC made its home in the Herskowitz Building in Oklahoma City. As Joiner had done with his son-in-law for Oklahoma Star, he now entreated another Ardmore man of good reputation, W. J. Newsom, to do the face-to-face stock sales. Joiner, for his part, began placing ads in Oklahoma newspapers far removed from any oil production. Under the banner MILLIONS IN OIL, his small ads read: "Do you want to know something of Oklahoma's wonderful oil fields and the millions made quickly in oil investments? Write for a booklet—it's free." The company boasted holdings of $100,000 and leases in 15 counties. It claimed 2,500 stockholders, with 90,000 shares sold to Ardmoreites. More likely than not, Joiner had sent subscribers an offer to convert their Oklahoma Star stock to White Rock stock for a small fee.

While A. D. Lloyd was still on Gorton's payroll as a geologist, he entered into a stock selling scheme with Joiner. In May 1917, the ad campaign began promoting stock in the Home Refining Company, capitalized at $1 million. Joiner was first V. P. and Lloyd, of course, was the chemist and geologist. The refinery was actually built, though not by Joiner and Lloyd. Instead of the 10,000 barrel a day plant advertised, it barely eked out 4,000 barrels as a skimming plant when operational and soon went bankrupt. Our East Texas boys didn't hang around long enough to see that happen.

By December of 1917, Joiner and Lloyd were both through with the Home Refining deal and Joiner was distracted by personal business back in Ardmore, where Lydia and most of his children still lived. In 1907, his son Bert had been tried for the murder of his girlfriend and was acquitted. But he refused to shy away from alcohol and reckless behavior. Now, a decade later, Bert had been shot and killed. He was 32. Two local men, Buck and Doc Criner, were charged with his killing. On his deathbed, Bert swore that it was Doc Criner who shot him. Criner was seen kicking Joiner in the head after the gunshot was heard. Testimony from eyewitnesses, however, revealed that Bert had been intoxicated and drew a gun on one of the Criner boys, neither of whom were armed. Bert lost control of the weapon, which fell to the floor and discharged, shooting him in the abdomen. Doc Criner admitted that in the adrenaline rush of the moment, he had kicked Joiner in the head several times to make sure he couldn't get at his gun again. Both Criner boys were acquitted in February of 1918. After burying his son, Dad Joiner returned to work.

DOC:
WWI to East Texas

Doc Lloyd meanwhile, in late 1917, took to writing sensational reports and geological surveys for Wilwell Oil & Gas's stock subscription offering. Their full-page newspaper ads featured testimony from their geologist (Lloyd) about their excellent locations, as well as photos purporting to be of their equipment and one of their wells. All one needed to do to cash in on the quick money when Wilwell struck it big was to clip and return their "prosperity coupon" with one's money.

The memoirs of George Gorton's son, who came to Oklahoma as an impressionable teenage boy, reveal much about the character of A. D. Lloyd during this period. From the younger Gorton's 1975 book, *"A Big-ass Boy" in the Oil Fields*, we learn that Lloyd began the lies about being a soldier of fortune in Mexico, feuding in his "home" state of Kentucky, etc. around 1917. He further claimed that he had graduated medical school but never practiced medicine. Gorton described Lloyd as "utterly fearless, a hero of wars and fighting who could never adjust to a peaceful existence."

Young George's father frequently sent him on surveying expeditions with Lloyd, who taught the teen how to handle a gun. To test the young man's skills, Lloyd would drive them to back alleys and mean streets around boomtowns where hold-ups were common. Once in El Dorado, Arkansas, Lloyd and the young Gorton were surprised when a pair of armed men stepped out of an alleyway and approached their car. As he had been taught by Lloyd, Gorton fired. Lloyd did, too. They drove off, leaving the two men to die because, ironically, Lloyd's mantra regarding criminals was "kill them and get rid of them." Pity, the justice system didn't treat him accordingly when his number came up later.

In January of 1918, the elder George Gorton took a trip back to Wisconsin to witness the dedication of the new Sunday school at the First Baptist Church of Racine, which bore the name Gorton Hall since he had provided the funding. He asked his pal and resident geologist, A. D. Lloyd to accompany him north. Lloyd took advantage of the paid vacation to make some new contacts and explore the Racine area for oil investors. George Gorton, who had a well-established name there, had certainly done well in raising capital through his three syndicates for legitimate drilling in Oklahoma. Lloyd would exploit the area for money…and much more.

"Colonel" George Montgomery, Lloyd's real estate friend from Arlington Heights, had been given the position of business manager of George Gorton's oil company, now called the Gorton Trust. After a disagreement with Gorton over money, Montgomery went his own way and landed in Amarillo, Texas, where he played poker, ran some stock scams and traded leases for a living. Montgomery put in a call to his buddy, Doc Lloyd, and told him the action was hot in Texas. Lloyd informed Boss Gorton that there was gas and potentially oil to be found in the panhandle and, naturally, the boss sent him and the younger Gorton down to survey the area on several occasions. The elder Gorton later leased thousands of acres on which to drill from the LS Ranch.

That summer, Lloyd prepared an exciting stock-selling survey for Great Western Oil & Refining Company for their acreage in the Pecos Valley, and declared that they would hit a 200-barrel per day producer. He made a final trip to Amarillo with young George Gorton just before Thanksgiving, then was off to Houston to visit May and the kids for a spell. 1919 was to be a very busy year for the big man.

From Houston, Lloyd made tracks for Mexico. He had some business down in Webb County for a new deal he was putting together and then just went on across the border. At Laredo, he applied for a passport in April, stating that he intended to cross the footbridge and visit most of northern Mexico, as well as some interior states to examine oil and mining properties there. He stated his legal name as A. D. Lloyd and his birthplace and that of his father as Breathitt County, Kentucky. The timing wasn't quite right for Lloyd to be taken seriously in Mexico, though. There was tumult in the government but not enough for his taste. He wasn't seeking to buy or lease land in Mexico; he was seeking a government paycheck and an official title for rendering his "professional" assistance in developing their oil and gas resources.

He was back in Fort Worth in about a month to nurture a new alliance with a name that is probably familiar to anyone who has driven through Dallas in the last few decades—Harry Hines. Before he was chairman of the Texas Highway Commission, Hines was an oil promoter and a bit of a wildcatter. When he created his Harry Hines Leasing & Producing Company in the spring of 1919, he made Adelbert D. Lloyd his vice president. His big advertisements touted all of the firsts that Lloyd claimed responsibility for—mapping this or that anticline (one of Lloyd's favorite words), discovery of the Cement field, discovery of the Medicine Mounds in Hardeman County and more. Hines advertised his stock salesman positions in papers across the Midwestern states seeking "producers with pep" to push his oil securities.

Lloyd's visit to Wisconsin with George Gorton the previous year had borne more fruit than just contacts with potential investors. In Grand Rapids, through a gullible immigrant investor named Julius Nelson, he had met the young Marie Nelson. Lloyd consulted with Mr. Nelson on the mineral possibilities of some land he owned and consulted with Miss Nelson on more personal matters. He managed to snag both Mr. Nelson's money and his daughter, securing the latter to get more of the former. Maintaining sporadic contact with the girl 22 years his junior over the course of the year, he finally made his move.

During a trip to Syracuse, New York with some other Hines characters in June 1919, Doc Lloyd wired Marie to come east to marry him immediately. She dutifully replied. The announcement of the surprise wedding in the bride's hometown newspaper (drawing on information obviously provided by Lloyd), said friends of the groom were shocked that he had married. "He was thought to be wedded to his calling," the article ran, but "he discovered his heart wasn't the granite he had supposed it to be." The news piece further trumped up Lloyd's credentials by saying he had prospected extensively in Alaska, India and South America and had discovered seven oil fields. The bride was described as having a "charming and pleasing personality." And just like that, A. D. Lloyd was a two-wife man once again.

The bride and groom had a weeks' honeymoon before they were to take up residence in his "large home in the fashionable subdivision of Arlington Heights." That isn't where they ended up. Marie stayed in Wisconsin with her parents while Lloyd attended to his various interests. In July, he accompanied the Gortons to Amarillo to select the locations for wells to be drilled on the LS Ranch. Meanwhile, the Harry Hines ad campaigns in Wisconsin ramped up immediately following the news of the nuptials. His doting wife served as his boots on the ground in Racine, to beam at the big showy ads that contained her husband's name and proclaim him to be a true oilman.

In September, the newly wed Lloyd saw his golden opportunity to seize upon the unrest in Mexico. He had followed the arc of Venustiano Carranza's presidency and concluded that Carranza would be very interested in the wealth that could be pulled out of the Mexican soil. He journeyed there and obtained an audience with the President by way of the governor of Coahuila. Because of the unrest and the peoples' displeasure with Carranza in general, Lloyd was provided with a cavalry escort while performing his "geological" survey for the Mexican government. He crossed the border back into Laredo on October 1, 1919 and went to retrieve his Wisconsin wife.

Again, the couple did not reside in Arlington Heights. It is probable that Lloyd no longer had a Lake Como home there and had indeed burned many bridges when he left the real estate business. He brought Marie instead to Wichita Falls so that she could see firsthand what a boomtown environment was like at Burkburnett. In November, he told her that a 3,500 barrel a day gusher had been named in her honor. She immediately wired the exciting news home to her sister in Racine. Her sister, in turn, immediately notified the newspaper, as Lloyd had hoped.

As the Roaring Twenties were ushered in, federal census takers greeted the postwar public with their myriad forms and questions. The picture of Lloyd's life as depicted by census data presents new mysteries with no reliable answers. Wife May was still in Houston with her 2 children, rooming with a Canadian shoe salesman and his two young sons. Lloyd himself is still at the Westbrook Hotel in Wichita Falls, but without his Wisconsin bride. Back in Wisconsin, her father, formerly a farmer, stated his occupation as that of an oil investor. (This would change back to "farmer" at the dawn of the next decade after Lloyd cleaned him out.)

The Nelsons have a surprising new addition to their family in 1920, just months after their daughter's marriage to A. D. Lloyd—they are listed as the grandparents of a child named Lila Lloyd, born in 1912. It is possible that she was a Doc Lloyd issue from a previous dalliance. She may have been a child that Marie or one of her sisters had out of wedlock, who Lloyd in his cavalier way offered to adopt, then promptly disappeared. You can almost hear him boldly declaring, "The oil field is no place for such a young babe. She'll have to stay here." She could have been born to stepdaughter Lillian Durham on the way out to Washington. Whatever her origins, it is certain that the Nelsons raised Lila as a dear grandchild, doted on her and sent her to good schools. Their daughter, Marie, at least publicly, did not show any interest in the child as her own Lila's origins remain a Lloydian mystery.

Mexican president Venustiano Carranza, with whom Lloyd conferred back in the fall of 1919, was assassinated on May 21, 1920. To keep his ties to Mexico, Doc forwarded on his survey findings to Carranza's successor, hoping that the new regime might be more interested in paying him to run oil operations than was Carranza's. In early June, Lloyd had returned to Fort Worth from Wichita Falls and filed his 40"x33" blue-line map, titled Plano General de la Zona Petrolifera de Mexico, for copyright protection, as well as his Recursos Minerales de Mexico. Alas, he still wasn't offered a gig in Mexico, so he kept the wolves at bay by writing more of the bogus reports for which he had become known among oil promoters. He spent much of the summer in New Mexico for the Kansas-New Mexico Oil Company, writing a report and selecting well locations. Kansas-New Mexico did drill two wells near Artesia, New Mexico, and one made a small showing of oil but was soon capped.

Doc Lloyd was called back to Oklahoma by the Gortons in January 1921. Two weeks after the discovery well came in at El Dorado, Arkansas, Mr. Gorton sent his pet geologist down to scout the area. He set up a small office there and awaited the arrival of Gorton's son. After mapping what he termed the Wilmington-Lawson Anticline, Lloyd left the young George Gorton alone to chase leases while Lloyd chased skirts, namely a local married woman who was willing to be his secretary and sleep with him occasionally. He also ran out to Lafayette County, about 65 miles from the El Dorado field, to select a location for the Arkansas-Texas Company, a genuine wildcat outfit. According to Gorton, Jr., Lloyd was in charge of their money while they were in El Dorado and frequently blew through it, leaving the boy high and dry while the good doctor drank and searched for a new scam promoter to attach himself to in booming El Dorado. He found the ideal partner in the person of Abner Davis.

By the summer of 1921, Lloyd was in on the slick promotions dubbed "The Abner Davis System Petroleum

Products." Davis employed a very creative copywriter for his full-page ads and had located an engineer by the name of Blackburn who had invented a gizmo called the Radioscope, which he believed could accurately locate oil-bearing formations. Throughout the younger Gorton's memoirs, he shares how Dr. Lloyd's life had a tremendous impact on his own life. Lloyd's word was gospel. In the chapter on oil promoters, Abner Davis in particular is described as a very unsavory character and con artist that Gorton encountered in El Dorado. Clearly Mr. Gorton had not seen all of the enormous Abner Davis "Millionaire Club" ads running in Texas, Oklahoma and elsewhere, touting the sainted Doc Lloyd as Davis' geologist "and practical oil field man without peer in this country."

This wasn't Abner's first rodeo, either. Operating out of Fort Worth in 1919, Abner Davis had promoted his First National Refineries scheme, the basis of which was essentially this: Buying in at $100 or more, a man was helping to fund the tweaking of the processes necessary to create one-man refining operations in which each man would refine 50 barrels of oil per day then sell the finished gasoline out the back door of the refinery. He continued with his promotions until the mid-1920s when he was charged with mail fraud. He somehow managed to get out from under the charge and Blackburn, the Radioscope inventor, went to prison instead.

It is apparent that Lloyd also bragged about his medical career while in the fray of the oil fields. Gorton notes in his memoirs that soon after Lloyd hung out his shingle at their El Dorado office, which simply read "Dr. A. D. Lloyd," many familiar faces from other boomtowns came in seeking medical attention from the sainted pseudo-geologist.

Perhaps Lloyd was still dabbling in the medical arts on the side for whiskey money.

After one of Lloyd's absences from the Arkansas office, during which he was probably just up the road splitting the stock sale profits with Abner Davis, he returned with a fanciful proposition for the impressionable Gorton and another young gentleman who helped out in the office. Lloyd had just come from Mexico, he said, and a group of his friends there desired him to head up a revolution. They would start in Laredo with guns and supplies from the United States, then drive into Nuevo Leon and Coahuila, picking up revolutionary followers as they went. If he headed the revolution as a general, Lloyd wondered, would the two boys be his aides? What nineteen year old male could say no to such a proposition? Lloyd naturally reconsidered the proposition over the next several weeks and decided against overthrowing the Mexican government, much to the chagrin of the Gorton boy.

With the situation in El Dorado no longer requiring his attention, in 1922 Lloyd drifted back to Oklahoma. He wrote his usual reports about imaginary anticlines to help promoters sell stock and he surveyed lease blocks for wildcat operations. By the end of the year, he had issued a press release to *Petroleum Age* that he was dragging up from Fort Worth and moving to Oklahoma City. He even made a trip to Houston to finally retrieve wife May and their two children. Now 52 years old, he was finally ready to actually domicile with one of his wives. At least it appears that way on the surface of things. But he wasn't one to stay at home for very long. What Lloyd lacked in real training or ability to locate oil, he made up for in his ability to hunt up greener pastures. He had weaseled his way into a "major interest" in some old abandoned mines near Lampazos in Nuevo Leon.

In June of 1923, Lloyd and 20 year old George Gorton went down to check out the properties. Gorton's father did not necessarily approve of this adventure, but Lloyd's offer to the young man of a one-eighth interest in the mines was enough to get him moving. Most of the workings of the mines had been destroyed by the infighting in Mexico in the past decade, so Lloyd's lease hadn't been difficult to procure. After a short visit, the party headed back to Oklahoma, leaving their hired engineer in charge of rebuilding the mines. It is likely that Lloyd did not have the money to fully fund the venture. But when Gorton inquired of him as to why the mine endeavor had ceased, Lloyd replied with the usual air of class that hovered around him like a thick black cloud of biting flies. He stated that he and the engineer had a disagreement which ended in Lloyd getting hammered and taking a "shit in the middle of the plaza in Lampazos," and being run out of town.

In the intervening years between his departure from the auspices (and payroll) of the Gortons and the East Texas discovery, Lloyd laid low and moved May and the kids back to Fort Worth. He ran a mail order stock promotion advertising "a mountain of gold" in Mexico to supplement his geological survey income. His stepson, George R. "Tex" Lloyd, as much a ham as his stepfather, toured the country as an accident insurance salesman with one hell of a gimmick. Actually, he had so many gimmicks that it is difficult to choose which ones to include here. He started on the vaudeville circuit as a young man. After taking the insurance salesman post, he used Lloydian showmanship to make a name for himself. Much like his stepfather did in the 1890s, he'd roll into town and put on a show with one or more of his talents. Then he'd sell insurance. He was a trick roller skater, a crack shot, an expert ping pong player, comedian, singer and dancer, played many music instruments, fenced, boxed, dabbled in magic and ventriloquism, and yodeled. He claimed to have sold more accident insurance policies than any other man in the world. He traveled extensively throughout Texas peddling his wares in boomtowns. Meanwhile, his sister Lillian had settled in Fort Worth and was exhibiting her oil paintings and writing poetry.

What transpired between A. D. and Marie Nelson Lloyd, a.k.a. Wisconsin Wife, is not known exactly. What is known, however, is that by 1928, whether because she believed her husband dead or just wished it were so, she was presenting herself as a widow. Back in Racine, when she was mentioned in the newspaper, her name was given as Mrs. Marie Lloyd (the proper form for a widow), rather than Mrs. A. D. Lloyd (the proper form for a married woman.) Doc likely sent her home to her parents when he began moving around. When her father's investment money dried up, so, too, might their marriage have followed suit. It would not have been out of character for Lloyd to have one of his buddies send her a telegram with the news of his "death" and advice that there would be no funeral to attend since he'd been blown to smithereens in a well explosion. Whether he played dead or the Nelson family just chose to present him as dead we may never know.

Dad Joiner sought out his buddy Doc Lloyd in 1927 after he'd finished leasing up land in East Texas to serve as the block for his East Texas lease share promotion. A lawsuit filed in Cherokee County, though, indicates that Doc

Lloyd was doing more than just consulting with Joiner and was in East Texas earlier than has been previously thought. In a suit filed in May of that year, one J. Z. Thomas alleged that he had owned in fee simple 160 acres in Cherokee County, and had occupied and made improvements to the land for more than a decade. The suit contends that Lloyd, on January 4, 1927, "entered upon force and arms" and dispossessed Thomas of his land without paying any kind of rental on it. Thomas was suing for $100 per year rental on the property.

Seeing the success Joiner was having selling syndicate stock and leases to the locals based on the false geological report he had written, Lloyd finally took the plunge and created Lloyd Oil Corporation, based in Fort Worth and incorporated in Texas and Louisiana. He incorporated Lloyd Pipeline Company likewise. Then he organized the Lloyd Oil Corporation of Delaware to act as a holding company for all of the bits and pieces and to act as an additional buffer in case he found himself in a legal pickle.

Smarting from a mail fraud order issued against him in December of 1928 for the Mexican gold mine promotion, he needed to switch to something he could push locally. There were so many bogus stock offers pouring out of Fort Worth at the time, that postal authorities began to sit up and take notice. Mail moving out of Fort Worth was subject to inspection. And in the run-up to the Daisy Bradford No. 3 discovery well, Lloyd found himself constantly in hot water. Following complaints from scorned investors, in 1929 the National Better Business Bureau issued two national bulletins warning against Lloyd and any securities he was offering. The Texas Blue Sky Commission denied his application to sell Lloyd Oil stock in the state of Texas.

Still he persisted, planting stories in the *Dallas Morning News* about the "legitimate" activities of Lloyd Oil. He even named an anticline after himself in Brown County.

Unable to sell Lloyd Oil stock locally with his usual brashness, and being watched by postal authorities, Adelbert D. Lloyd switched gears entirely in August 1929. While Dad Joiner was in East Texas and Dallas still flogging the East Texas lease promotions, Lloyd stepped away from the oil scene. He organized the North American Lloyd Aircraft Corporation. Although he needed to be discreet to avoid trouble, he still could not resist the urge to put his name on the new venture. Stock certificates were printed up and letters went out to the sucker lists he'd built over the years. The idea behind the scheme was an airline from Canada to South America which, of course, had to make fuel stops in between. Those "in between" places stood to gain by having the air traffic stop there, and it was those very places that Lloyd visited while trying to promote his latest flight of fancy. The venture, suffice it to say, wasn't on the level and never got off the ground.

O all you young ladies who live in the flat
   Beware of the cowboy who wears a big hat
For he'll rope you and throw you
   and when you're well tied,
He'll jump on your bronco and there he will ride.
   He'll not ride you fast or ride you slow
But he'll jump up and down on your bucking bronco.

*- Ribald tune hummed and sung by A. D. Lloyd during his
time in Amarillo with young George Gorton. Original
author unknown.*

Doc's Infamous Geological Report on East Texas

[Were it not for A. D. Lloyd's "scientific" report on Dad Joiner's lease block in East Texas, he likely wouldn't have become a blip on the historical radar of Texas. And without it, ol' Dad may not have sold enough lease shares to have his name known to us either. Although it appears in other histories, it is included again here because of its significance as the report that put Lloyd on the map.]

Geological, Topographical and Petroliferous Survey, Portion of Rusk County, Texas, Made for C. M. Joiner by A. D. Lloyd, Geologist & Petroleum Engineer

TOPOGRAPHY

The Overton Anticline is located in the Juan Ximinez, Isaac G. Parker, R. W. Smith, Geo. W. Guthrie, Bennett Smith, T. J. Moore and M. J. Prue original surveys, or land grants which have been subdivided into small tracts belonging to the present numerous owners, reference to which is hereby made to Rusk County ownership map as compiled by C. M. Joiner.

The area covered by this survey is marked by the presence of high-rounded, steep-sided, sand-covered hills, with narrow streams traversing narrow, sandy valleys. The Overton Anticline is located on a ridge from which the streams flow

North into the Sabine River, and South into the Neches River. The smaller streams are fed with springs and seldom become entirely dry. The larger streams are deep and sluggish and have the general appearance of Gulf Coast Bayous. The higher hills are covered with Pine, and the valleys are marked by the presence of Live Oak, Post Oak, Gum, Maple and Willow trees. The streams and forests will furnish ample water and fuel for developing purposes. The area is traversed by hard-surfaced highways with laterals that are generally graded and passable for heavy equipment.

<u>STRATIGRAPHY</u>

The formations deposited in the regions surrounding the Overton Anticline belong to the Wilcox Eocene (Claiborne Group). The formations dip South from the Overton Anticline which is marked by the presence of the Yegua Formation until within twenty-five miles distance from the Overton Anticline; the Cook Mountain and Mount Selman formations form the surface sediments. On the North and East, Cook Mountain is the highest formation encountered within twenty-five miles. On the North and West nothing higher than the Cook Mountain formation was identified during this survey.

The areas covered by Rusk, Cherokee and portions of Smith Counties, Texas, have suffered greater distortion generally than any other similar-sized area in East Texas or Louisiana with the exception of that area covered by the Sabine Uplift in N.W. Louisiana; and there are no other areas of similar magnitude where the position of the strata can be interpreted by engineers with the degree of certainty that is possible in this area. This enables structural conditions to be defined with a degree of certainty quite unusual for Delta Structures. During the time now elapsed since the bringing in of the Humble Oil Co.'s No. 1 Discovery Well at Carey Lake, Jacksonville, Cherokee County, Texas, this regional high has attracted the attention of all the major oil companies as well as the important independent operators, and large blocks of acreage have been leased by the Roxana, Shell Oil Co., Magnolia Oil Co., Texas Co., and the Humphries Co. The above blocks are all in the counties mentioned, and the Overton Anticline (C. M. Joiner Block for whom this report is made) is located in the center of the above leased blocks. Much of the land lying between the above named blocks has been leased by independent companies who paid unusual prices for same.

Faults, Folds and Dips

The Overton Anticline is located near a Saline Dome that has generally disturbed the strata of the region. This dome is marked on the surface by a Salt Lake, or Marsh, in the N.E. Corner of a 104.75-acre tract now belonging to the Mayfield Co., and by the presence of a Salt Crystal that forms on the bottom of a well dug for water on the N.E. Corner of a one-hundred acre tract belonging to Calvin Young, and by the presence of a Salt Lick on the S.E. Corner of a 74.5-acre tract belonging to J. A. Birdwell. All the above is in the M. J. Prue Headright Survey. The above occurrence of a salt in the earth and water forms a triangle with a lateral line of about one mile, and is from two to three miles from the apex of the Overton Structure.

The Overton Anticline, while lying close to a Saline Dome, is not a Saline Dome Structure, but is a Faulted Anticline traversed by an East and West Fault extending in a general direction sixty-eight degrees North by West. This Fault Course is crossed on the highway between Arp and Overton, where it is marked by a well-defined arch, and on the easterly extension it becomes the arch in the city of Henderson, county seat of Rusk County. This fault may be traced for a considerable distance in both East and West Extensions. On the East it can be traced as far as the Shelby County Oil Fields; on the West across Smith and Van Zandt Counties. This anticline is also traversed by a fault extending in a general direction North twenty-two degrees East. The North Extension of the fault would traverse the Caddo Oil Pool in the region of the Pine Island Saline; the South Extension traverses the Humble Oil Co., Cherokee County Saline where the No. 1 Discovery Well is located.

The red, white, blue and brown stratified sandstones and shales are sufficiently bedded and consolidated to enable clinometrical measuring of the dips that have been recently exposed by the cuts made in grading the highways of this region, and a great number of these dips have thus been exposed at numerous places where they develop arching, doming, and possible faulting, and the structures that may be developed by these exposures justify, in the author's estimation, the present activity displayed in this district. The Aerial Geology of the Overton Anticline is well marked and developed by the well-exposed strata dipping in all the cardinal directions from the apex of the anticline. These dips have enabled the petroleum engineers to place the seismograph in the most advantageous position to receive the concusionare vibrations. Under these favorable conditions thousands of registrations have been made on and around this block by major companies.

Anticline and Depth to Pay

The Overton Anticline displacement has probably resulted from the intrusion of the Saline Core to the southwest, and the upthrow of the traversing fault is found on the south and east quadrants and the downthrow on the North and East of the intersection of faults.

There is a drag toward the apex in the Conture Lines of the N. W. and S. W. Quadrants, which are the larger quadrants of the structure. There are two outlying small, slightly elevated Dome Areas lying just off the Central Drag of the Conture Lines of these two major quadrants. These two outliers will be important producers. The entire structural condition forms a geometrical arrangement that is exceptionally balanced, which justifies the conclusion that the accumulation of oil and gas will be of unusual importance. The author expects some very large wells to be developed on this structure.

The location of the C. M. Joiner Well No. 1, on the Daisy Bradford Tract near Johnson Creek will spud in the top of the Yegua Formation on the base of the Cook Mountain. Fossil leaves and plants of undoubtable Yegua were obtained at a depth of twenty-five feet from a water well dug on the H. M. Cooper thirty-five acre tract which showed similar fossilization. Salenite Crystals have been found in several water wells and reported as Mica by the land owners. Many of the water wells have water with unusual taste that accompanies the presence of Alum and Gypsum. These salts and fossil remains, and the dips which enable the strata to be traced for great lateral distances substantiates the conclusion that this well will begin in the top of the Yegua formation.

The log of the Rucker Well, drilled on the Artic Wright Land located on the Parmelia Chisum Survey about five miles North by East from Henderson, and eight miles East by North from C. M. Joiner Location No. 1 encounters the top of the Austin Creek Chalk at 1,497 feet. The strata between the Rucker Well, located in a syncline, and the C. M. Joiner Location are sufficiently exposed and continuously connected to correlate between the two wells, and such correlation shows that the Rucker Well cut 175 feet plus what will be cut by the Joiner Well. This also substantiates the dip to the Northeast.

Base of Austin Chalk in Rucker Well	2,500 feet
Top of gas sand in Rucker Well	3,270 feet
Base of Austin Chalk in Humble Cherokee Well	3,004 feet
Top of Oil Sand in Humble Cherokee Well	3,843 feet
Which checks the formations in these wells within	69 feet
Base of the Austin Chalk in the Rucker Well	2,500 feet
Top of Limestone in Rucker Well	500 feet
Base if Austin Chalk in Humble Cherokee Well	3,004 feet
Top of Limestone in Humble Cherokee Well	3,706 feet

which correlates the formation and pay sands in the two wells to within a few feet. The C. M. Joiner Well should encounter pay in the stratum from which the Rucker Well produced gas at a depth of 3,230 to 3,270, at a depth of 3,055 after deducting 175 feet of strata cut by the Rucker Well, which will not be cut at the Joiner Location.

These correlates show the thickness of the sediments are continuous and uniform over an area from the Humble Oil Co., Cherokee County Well to the Joiner Location in Rusk County.

The Producing Oil Sand

The producing zone in both the Rucker and Humble occurs at the zone of the first limestone and beneath the Eagle Ford Shales, which zone is classified as the Woodbine Sand. There is a possibility of picking up a pay in the Nacatosh Sand at a shallower depth. The Rucker Well picked up a stray shallow sand at a depth of from 577 to 604 feet. This pay horizon will be encountered at a depth of 402 to 429 feet. The Trinity Sands should be encountered at a depth of 4,200 feet. These producing oil and gas sands in the fields now developed in the region surrounding the Joiner Well yield large gushers.

Location Joiner Well No. 1 is made on the Daisy Bradford Tract in the Juan Ximinez Headright Survey.

DAD:
WWI to East Texas

After burying his son, Bert, in 1917, Columbus Joiner continued to sell stock in his White Rock Oil concern. He and Lloyd may have stumbled upon each other now and again, but Lloyd was not held out as Joiner's geologist during any of White Rock's stock promotions. By the end of 1919, many of the lots Joiner owned in the Joiner City area outside of Ardmore were being put on the auction block for past due property taxes dating back to 1915, for sums less than $2 per lot. One of his first corporations, Joiner City Oil & Gas, was killed by the state of Oklahoma for nonpayment of taxes. It was clear that he had milked Oklahoma for all he could and it was time to move on. He still made Oklahoma City home, but activity there would no longer be his livelihood.

In November 1919, Dad Joiner began again (again) in Texas. He hadn't gotten rich in Tennessee. The fresh start in Indian Territory didn't do the trick. And moving to Oklahoma City didn't change his luck, as he saw it. So he set his sights on Texas. The new mask donned by Joiner for the Texas crowd first bore the name Joiner Investment Company. But it didn't take Dad long to figure out that if he wanted to hide from his creditors and from postal fraud inspectors, it might be best if his name wasn't in the company name emblazoned across the letterhead of his promotional pieces. So, in December, he gave his Texas promotion a new moniker, one more likely to engender a sense of community and patriotism. He called it the Citizens Lease Syndicate and rolled out ads selling lease shares. For one amount, an investor would receive a share in a lease in the county of Joiner's choice. But for six times the money, they'd have a stake in all six counties in which Joiner claimed to hold leases, although the ads only listed five—Rusk, Panola, Shelby, Cass and Nacogdoches. In 1919, the ads included Smith County, but it was phased out as 1920 began. He promoted under the Citizens name up until the time of the East Texas discovery well.

In May of 1920, the Joiner clan—Lydia and most of the children were still in Ardmore, taking in boarders to make ends meet—suffered a loss. Joiner's sister, Mamie Simms, lost her husband in a tragic and bizarre shooting in Idabel, Oklahoma. John Simms was a police sergeant working the night desk on May 20, when two local and rather inebriated toughs pulled up in front of the station. When Simms stepped out to see what the men wanted, they unleashed a very large, angry opossum from the car. Simms directed them to remove the animal, but they refused. The more sober of the two men got out of the car and helped another officer corral the creature, while the more toasted of the two sped off. He made the block and pulled back up to the station, where an argument ensued between he and Sergeant Simms. The sloshed man pulled a .32 caliber automatic pistol and Simms approached the car in an attempt to disarm the situation and the man. Nine shots were reportedly fired. Five shots told in Simms' body. Lydia Joiner, Dad's wife, went to her sister-in-law's aid but no mention was made of C. M. Joiner attending the services of his sister's beloved husband.

Although operating under the name Citizens Lease Syndicate for the better part of two years in a mail order capacity without doing much actual drilling, Joiner did not feel the heat from investors until 1922. He sent a press release to *Petroleum Age* in January of that year which simply read, "The Joiner Investment Company...has changed its name to the Citizens Lease Syndicate and plans activities." Activities, indeed! Lots of sales letters to sucker lists of Ardmore folks and East Texas farmers, as well as widows and doctors in other states definitely constituted "activities" for Dad Joiner.

It is common knowledge that Joiner began his activities in East Texas around 1925. He charmed his way into Daisy Bradford's place in 1927. The lease stipulated that a well must be sunk in order for it to remain in effect. But Joiner was advertising the spudding of test well six miles south of Overton (the vicinity of Daisy's farm) as early as October 1922, a full three years before history puts him there. Whether the well was truly begun or if the press release was just a puff piece is lost to history, but the fact remains that he had his eye on barren East Texas as fertile ground for local lease sales well before 1925. It is doubtful that he believed that East Texas was "floating on a sea of oil," but Joiner's low-key Bible-quoting charm earned him some desperate and loyal local support early on. He was listed in the 1925 Tyler, Texas directory, along with his youngest son, Verne Joiner. Both John and Verne would join their father in Texas, leaving Lydia to fend for herself in Ardmore.

1923 to 1928 found Joiner ferrying back and forth between Dallas and East Texas, taking care of his promotions. He visited his daughters in Ardmore on a couple of occasions, bringing them a trinket or two and contributing to the family finances in a meager fashion. He had also hired on a 19 year old secretary to help manage the mountain of paperwork generated by his schemes, Miss Dea England, a Dallas native from a working class family.

By the time the first two wells on the Bradford farm failed and Ed Laster was making an earnest effort to bring in the third well, Dad was writing letters to the editor of the *Dallas Morning News* of a peculiar bent. Some of his missives were published and one of August 23, 1929 read like something out of the late 1960s. Dallas was trying to cope with the drought and crop failures and the run-up to the Great Depression and Joiner had a solution to it all: Hugs and kisses. One imagines him writing this en route to Woodstock in a 1964 Volkswagen Microbus, rather than weeks before Black Tuesday:

SAYS LOVE IS PRIME NEED OF DALLAS CITIZENS RIGHT NOW

Much has been said and written as to the thing most needed in Dallas. As love is the greatest force in the world, the opening avenue for the successful accomplishment of good things, why not make Dallas the City of Love, blending every interest into a flower of love, each and everyone becoming a torch bearer? The effulgent light of love reflected from every soul will drive from our city all feelings of discontent, every feeling of friction and unrest, and make way for the love that knows no bounds—a love that will cause the heart to beat in unison in the upbuilding of Dallas and for the true happiness of all her people.

There are many heavy hearts in Dallas that beat as true as any hearts in the world—men, women and children who, by mismanagement and otherwise, have lost their health, their wealth, their prestige and their happiness. They are crying for love. Love would heal every wound, If we could all say feelingly, "Courage my brother; courage my sister. We will throw around you our impenetrable shields of love and will help you in your efforts to rise together," Dallas would be known throughout the world as the great City of Love.

What Dallas and the rest of the country needed, of course, was money and prosperity but Dad Joiner was of a romantic ilk and enjoyed showing off his pseudo-intellectual, reverent, soft-spoken nature which served to impress certain citizens of Texas. He very much wanted to be thought of as an intellectual man and seemed hell bent on being a man of letters. As Ed Laster was trying to drill the discovery well, Dad Joiner was up in Dallas in his hotel room feverishly writing. Not just promotional material to be sent to his sucker lists, either. He was working on a small book that he hoped would be picked up by bookstores and sold. He was dipping a geriatric toe into the world of publishing and paid old Adolph Geyer at Geyer's Printing to whip up an unknown quantity of his masterpiece. But the words on the page were from a different era, woefully out of fashion and not particularly well written. He delivered signed copies to booksellers in Dallas and even mailed a copy to the drilling site on Daisy Bradford's farm. It's a safe bet to take that he did not mail a copy to his wife, Lydia, back in Ardmore.

Parts are very Victorian in their way, while other parts seem like they'd blend in if put to music and sung by Conway Twitty circa 1975. Without further ado, here is C. M. Joiner's 1929 ode, *Love Letter to Heloise*, no doubt inspired by his devoted secretary, 50 years his junior. Reader beware—heaving bosoms ahead:

I tried to strip the veil of law and custom from the world to lay bare the naked, pure and beautiful of all nature clothed with a beautiful smile which is the spirit of all form. I substituted the elysian fields for the Garden of Eden that I might blend it with the awakening civilization as described in the Lovers' dawn of the first day when they found on awakening that love is spontaneous, and that they were only the central figures in the great drama of life. With this understood by the reader, a much more liberal view may be taken of the setting as the picture is purely an imaginative one, but holding true to Nature and Nature's God—Love.

My Own Darling Heloise:

My heart warms to every breath of you. I can never feel cast down to earth with the thought of you awakening the sweetest hopes; the highest aspirations.

I have read many of the gems of Ancient and Renascent Literature. I have roamed the realms of art and viewed with tenderness and love the verdant hills, the valleys and waterfalls. I have viewed with wonder the tempestuous Atlantic and the serene and placid Pacific. I have with an open mind and heart seen the glorious dawn as the great orb of day would throw its warm stream of life and light across the earth. I have seen the glorious sunset on the Western prairie. I have heard the native birds and the various migratory winged songsters as they wended their way to other climes. I have, with an open heart and uncovered head, stood beneath the canopy of Heaven and viewed with awe the glories of starry firmament. But as glorious as is the panorama just described, it does not fathom the depth of my soul like being with you, sweetly enfolded in your loving embrace and looking into the windows of your soul to see reflected there all the beauties of earth and sea. It is then that my heart vibrates with the central luminary of God's noblest handiwork—my Heloise.

All extremes cure themselves except love; in that, the only infallible remedy is a true and loving mate. "The Leaves of Grass," by Walt Whitman, and "The Sower," by Millet, embody the finest things in literature and art, but, Dear, they only fertilize the mind and heart for the great awakening that lies dormant in every human heart until touched by divine love. Then, and not until then, is all the beauty of Heaven and earth revealed to the human heart. It is then that one can hear the murmur of the streams, the songs of the birds, the onrush of the storms and the breaking away of the dark thunderbolts, revealing a silver lining behind every cloud.

When you are gone I am lost in an impenetrable gloom, "God moves in a mysterious way, His wonders to perform." He only can be my solace when you are away. It is then that I can hear the howling of the tempest and the imprisoned spirits, and feel that I am standing on the windswept earth between two eternities. It is then that my heart appears to be dead to all the awakening glories of earth and sea, then I hear the melody of your sweet voice, which is to me sweeter than the songs of the nightingale when the fragrance of sweet-scented flowers are floating in the twilight.

Oh, what melting delight in your kisses. If a butterfly could sip the nectar from your lips it would be intoxicated and would hide behind a flower, or would fly away into the glorious sunlight while your voice would rise to blend with the voices of the angels.

I am now in the twilight looking for my darling. I can feel the heaving of her snow-white bosom, and feel her breath and hear her sweet voice. I am now wafted with her into Elysian fields. We are happy. We are whispering to each other our undying love and devotion. The sun is sinking behind the Western horizon as the evening shadows are lengthening. We are clasped in a loving embrace. I can hear the angels' wings as they hover over us singing their songs of approval of our divine love. The earth is covered with moss-covered verdure. Here we repose as we love and love. We are alone. This is the supremest moment of bliss; "The moment of bliss not too dearly bought if purchased by death." We now fall asleep in each other's embrace only to be awakened by the lowing of the herd, the tolling of the bells, and the warbling of the winged songsters as they are mating as we have done. Now the awakening of a new life sweetly blended by the power of love. We are one; our love is for eternity. We are supremely happy, and bless every element that contributed to the sweet blending of our lives.

In conclusion, dear Heloise, I wish the melody of my language could be commensurate with the message of my heart, but the lofty conception must lie dormant for lack of suitable language—a language not found in our mother tongue—so the message can only be conveyed by the sweet harmony of Nature; in the breeze, the invisible messenger, like the breath of the tropical winds, gathering every sound and wafting it across every hill and plain.

I hope these words, though crudely expressed, may fall upon your heart as does the dewdrops that refresh and sweeten every flower. You are the flower of my heart; the master soul.

Operating out of his Dallas hotels trying to trade leases and stocks, Joiner always claimed to be broke. Spending his money on book printing probably didn't help the situation any. He had the occasional assistance of his eldest son, John, and Verne was also in on the action helping to sell lease shares in Texas. Halbouty and Clark tell us that Joiner's syndicate certificates were often swapped as currency in Overton because there was so much of his paper floating around. But when Laster pulled an oil-rich core from the third well and called Joiner advising him to come down and come quickly, no doubt Joiner panicked.

As Ed Laster got closer to bringing the well in on Daisy Bradford's farm in September 1930, after four months of downtime, Dad Joiner remained in Dallas, still pushing stock certificates and lease shares to keep money in his pocket, but losing his mind over how many investors might try to lynch him if oil actually came in. If oil in paying quantities was found, he'd owe so very many people a fractional interest in it. Since he had oversold the shares, he was potentially in hot water. What the world didn't need just then was more oil and prices were low. He was in a bind.

Meanwhile, Doc Lloyd was 150 miles away working a deal near Pilot Point in the Denton area. After Laster convinced Joiner to show up and see the state of things, Joiner informed him that his old pal Doc Lloyd would be arriving to complete the well. This, of course, made little sense.

Laster was a skilled driller who had worked for months with deteriorating pipe and inadequate equipment. He took umbrage to the idea of another coming in to finish what he had worked so hard to accomplish. When Lloyd showed up at the well site, there was an instant clash of personalities. Lloyd, with his take-charge personality, insisted that his crew would work the day shift and use their own techniques to bring in the well. Laster was relegated to the night shift, which was problematic considering there was no money for lighting.

After the encouraging drill stem test and Lloyd's arrival, the "eminent geologist" immediately began giving interviews discussing his past with Joiner, how they'd brought in gushers and opened four fields together. He gave his usual rosy speculation on how big and wide the field would be, and how it was to be of "major importance." Of course *every* area that Lloyd was paid to survey was going to be of major importance. The difference this time was that he was right. Even an old, gutted clock is right twice a day.

Just before the latest lease on Daisy's land was about to expire, an event which would have saved Joiner much heartache, Lloyd got to Daisy. He told her that he knew of a new, better way to bring in the well. She was impressed by his scientific vocabulary and called in Ed Laster to consult. Presumably from interviews with Laster, Halbouty and Clark recounted the conversations between Daisy and Ed in *The Last Boom*. Daisy informed Laster that Lloyd would now be in charge and employ the new techniques. She wasn't fond of the man, but had faith in his jargon and confidence. She considered him a smart man. Laster told her that if Lloyd had his way, the hole would be junked, just as the two previous wells had been. After some thought,

Daisy put a stipulation into the new lease that Laster was to complete the well with the materials and methods he saw fit.

When Joiner saw the new lease agreement, he was crestfallen and somber. He accused Ed Laster of working against him. It has been interpreted over the decades that what Joiner meant was that if Ed completed the well, there would be no oil because Lloyd knew best. But after a study of Joiner's past as a promoter and *not* a wildcatter, that seems unlikely. Seeing the specter of lawsuits and royalty payments haunting his doorway, Joiner had called in Lloyd to junk the well. No oil, no heartache. He saw his future as a much brighter one if the well was scrapped as the last two had been. But Daisy Bradford wanted a well and, as we all know, she got it. People who witnessed the well blow in, amidst the carnival atmosphere of locals who had been praying for oil, said that Joiner went white as a sheet when it happened. And one needn't wonder why. The well was completed on October 5, 1930.

The oil fraternity wondered if the discovery was a fluke or the real deal. A young independent named H. L. Hunt drilled a well nearby and six other test wells were down by the end of October. Hunt's well made a poor showing, but he was convinced that neither he nor Joiner had drilled in the right places. He knew Joiner was financially hurting but wanted to see if the field would open to the west before making a move. Joiner, of course, did not attempt another well, even though he had the leases to drill upon and the confidence (and financial backing) of the people of East Texas. Less than a month after the discovery well came in, Hunt had already begun a 4" pipeline from the discovery well to the rail depot 2 miles away.

Almost instantly, cash-starved local investors began trying to cash in on their syndicate shares. But Joiner had no money. So the hailstorm of lawsuits began and Joiner retreated to Dallas once more. As the story goes, Joiner sold most of his interests to H. L. Hunt by the end of November. Even though Hunt didn't have the full $1.3 million to give Joiner upfront, he forked over $30,000 as a down payment, which was still more money than Joiner had ever seen in his lifetime. The remainder was to be paid as the oil came in...if it came in. Naturally, it did, and Joiner received payments from Hunt on time as promised.

After the sale to Hunt, Joiner and Lloyd gave a few interviews to the press, both together and separately. By November, A. D. Lloyd was conducting business with his newly formed Lloyd Oil Corporation. He had taken up residence again in Fort Worth with May and their two kids. His son Douglas, from his marriage to Anna Meeks, was working with him, proudly touting his father's geological acumen and calling himself a geological engineer. Lloyd was back in the business of selling stocks and planted stories that he was drilling or had leased acreage everywhere from Red River to Zapata County. Joiner, meanwhile, laid low and soaked up the limelight a bit. He now had more than just his quiet demeanor and Bible verses to endear him to people. He had a name to hang his oil promoter's hat on.

In February of 1931, Joiner was in Alto in Cherokee County giving a speech to a crowd of 50 landowners, trying to convince them that he could do for them what he did back in October in the "Joiner East Texas Field," as he preferred to call it. Everyone wanted in on the excitement that had been brought about by the discovery so it was easy for him to lease up a 9,000 acre block. The leases stipulated that Joiner would begin drilling within 90 days. He impressed the people that day by going out immediately and arranging for timbers to be cut to erect a derrick. But the well never went down. When the lease expired in April, an old chum was on hand to pick up where he left off. Doc Lloyd and his stepson, "Tex" Lloyd, showed up. Tex put on a fancy roller skating show for the people and his father gave a talk about how the people of Alto should pool their money and offer a bonus to the first company to bring in a well. He claimed to have drilled the first well in the Lloyd Anticline, the Breckenridge Field, the Sipe Springs Field and many others. The people of Alto raised $7,500 that day and the people of Rusk raised $10,000. Tex also sold many accident insurance policies, to boot. Lloyd called Joiner back in to work his magic on the people and collect the money. The well came up dry and the derrick eventually blew over in a drought-breaking rain. Thus ended the final known collaboration between Dad Joiner and Doc Lloyd.

Lloyd Oil Corporation logo as represented on the title page of the 1931 Granville Cubage promotional book. The company's motto was "Build Lloyd—and Lloyd will build you."

Doc After Daisy

On January 10, 1931, Doc Lloyd and sons, Tex and Doug, along with 500 to 1,500 people (depending upon which sensational report you read) and a brass band, gathered together to watch the Tex Lloyd Camp No. 1 in Rusk County come in. It provided an excellent photo opportunity for promotional materials and Lloyd was in need of a boost. The authorities were sniffing around. Preliminary investigations for mail fraud had begun. Years of double-dealing and swindling were catching up with him. He kept his head down and played everything close to the vest. He had promised investors for years that he would bring in a $100 million well, but instead he had spent every penny they sent him. Now the portly pseudo-scientist was rolling downhill and no amount of bluster and bravado could break his fall.

A scholarly-appearing book, a little smaller than this one, bound in gold-stamped leatherette covers and written by Arkansas school-teacher-turned-Lloydite, Granville Cubage, called *Oil, A Handbook for Reference: A Study of the Lloyd Oil Corporation of Fort Worth*, was published shortly after the Tex Lloyd well came in. It is interesting to note that Mr. Cubage had moved to Oklahoma and was billing himself as a geologist in 1940, when prior to that he had been an Arkansas teacher and school principal for decades. That was the power of Lloyd.

The book, sent with promotional letters to impress suckers with particularly deep pockets, gave a sensationally fictitious account of Lloyd's life (as provided by the "doctor" to the author), his philosophies on the origins of oil and, naturally, his philosophies on wealth and oil investing. Cubage's reverence of Lloyd's scientific prowess and his unerring belief in the stories Doc fed him demonstrate clearly the influence Lloyd's domineering character had on even the educated family man. Lloyd told Cubage he'd single-handedly discovered 31 oil fields and made him believe it. He fabricated an entire life story and career…Cubage bought it and then wrote about it. Then Lloyd used it to do what he did best—sell more Lloyd.

Production statistics and an explanation of drilling process added to the book to lend legitimacy to it are omitted here, but other portions of the book are given below under the subheads under which they appear in the original text:

STORY OF A MOUNTAIN BOY WHO PLAYED A MAN'S PART

It happened back in the 70's of the last century. The afternoon sun was beating down upon the rugged hills of old Breathitt county, Kentucky—"bloody Breathitt" they called it with right good reason. The rays of that sun fell upon the barrel of a rifle which, poised for action, lay in the notch of an oak log behind which its holder was concealed. A glance at the marksman whose keen and well-trained eye traversed the barrel would have brought a little start of surprise to any not familiar with the conditions of that feudist country.

A robust lad of only nine summers lay there, with fire in his eyes and an expression of determination on his face which would have done credit to the head of any family in the settlement—for that is what he was, the head of a family, at the age of nine, a post which he had now held for two years. And to be the head of a family in Breathitt county in those dark and turbulent times meant more than being a bread-winner whose duties were confined to bringing home the groceries on Saturday night. Other responsibilities of a far sterner nature devolved upon him, for he had fallen heir to his father's place in the clan warfare of the region, and he must carry on.

In this he was no amateur, in spite of his youthful years. He had been a crack shot since the age of five. Since the age of seven he had been an orphan, rendered so by the ill-fortune of a luckless feudal fight in which his father had been killed. Since then it had been one of his duties, imposed by fierce mountain custom no less than by his own conscience, to avenge that death by exacting the old penalty of an "eye for an eye, and a tooth for a tooth." Under their hard code it was desirable to kill the killer himself, but, failing that, any member of his tribe would suffice.

On the mountainside a bough moved and a face peered cautiously out from its hiding place. Instantly a gleam of hard blue light flashed from the eyes of the boy and his trigger finger did its work. At the crash of the gun-fire, old "Red-bone" John ducked and fell sprawling in the leaves. The rifle ball had plowed a neat furrow through his scalp which would leave him permanently marked, though it was not fatal.

From that day on the youngster was regarded as an initiate in the bloody doings of the county and must forever be on the lookout to protect his life from those who had struck down his father. And what a character it was to be played by a boy of such tender years and one who bore the rather flossy name of Adelbert—Adelbert Durham Lloyd.

The Lloyds were of the old stock of pre-Revolutionary Americans who had pushed their way back from the seacoast generation by generation until they had at last come to rest in this wild and passion-torn region. In common with many of their neighbors, they inherited as good blood as was ever brought to America, and they were feudists simply because the customs of the times imposed a duty on every man to be ready and willing at all times to protect himself and his own with whatever he had of strength and skill and bravery.

Around the fireside that night the anxious mother, ever alert for the protection of her brood, counselled with her family for the safety of her first-born. It was but the first of many anxious months for her until finally she determined upon action which would assure safety for him. It was an evidence of her lofty character that she did so in a constructive way and one that would lead to his development and greater usefulness throughout life.

The result of the mother's plan was that young A. D. Lloyd was taken to Cincinnati at the age of twelve and placed in school. This was no new experience for a member of the Lloyd family, but it was a novelty for the branch of it which lived in Breathitt county. Back along the line before the great migration, before the time when it had buried itself in the great wilderness, it had numbered educators and public leaders among its membership. Now it was to reassert itself and send out from its isolated mountain retreat one who would make an impress on the scientific thought of his age.

Out of the Bondage of Fear

Happy school days now open up for him, days of simple, childish play, days that were carefree for the first time in life. Should you ask him for the most vivid recollection of the period, the experience which impressed him most, he would tell you of the time when a new-found friend, a boy of his own age, invited him to go swimming.

"All right," said young Lloyd, "wait till I get my guns."

"What do you want with a gun?" asked his friend in astonishment.

"Why, I just thought they would come in handy in case of a fight," young Lloyd explained.

The new friend was quick to show him the absurdity of such an idea and almost for the first time young Lloyd went unarmed to a gathering of his fellow human beings. As he stood for a few moments on the bank and watched the carefree movements of these boys who had grown up in innocence under conditions so vastly different from his own, he somehow caught the contagion of their happiness.

With boyish impetuosity he began to disrobe, and as he laid aside his garments to make the plunge he also laid aside the garment of fear which had cramped him almost from the day of his birth. He plunged and swam and dived with the utmost abandon, for at last he had tasted the glorious freedom which others accepted as a matter of course. For the first time he understood the sweetness of life in a world where he was secure from bodily harm.

Making A Record

Athletic prowess was not encouraged in the schools of that day, else young Lloyd, with his bodily vigor and strong determination, would doubtless have hung up a string of records. For scholastic work it might be surmised that he had meager preparation, but by earnestness and industry he made up his deficiencies and within a period of eight years we find this mountain lad finishing the public school course and the course of study of American Eclectic Medical College where he was graduated with high honors at the age of twenty years.

Although he graduated with the degree of Doctor of Medicine, his major study was chemistry and he directed his further studies along that line instead of the practice of medicine.

First, he spent a year as a professor of chemistry at his alma mater. While pursuing this work, he became much interested in assaying ores. It happened that about this time the transcontinental railroads were finished and the Pacific Northwest began to beckon to the ambitious. Accordingly, Dr. A. D. Lloyd, late of the faculty of American Eclectic Medical College, hung out his shingle as an assayist, in a little town in northern Idaho.

Turning Prospector

All these years his love of the mountains had lain dormant in his blood, and now that he was back among nature's scenes once more he could no longer resist her lure. In his work as an assayist he came in daily contact with the grizzled prospectors of that region, and it was not long until he had caught their spirit of optimism and found himself taking to the trail with pack mule and gold pan.

It was a life that required patience, and perseverance, to say nothing of courage, but after a few years came a time when his efforts were rewarded. On a wild and desolate point of the mountain, near the present town of Hailey, Idaho, he was plodding along behind his pack mules, when he suddenly heard his mule's hoof strike gravel. Ever alert for such things he brought his little cavalcade to a halt. Yes, it was a gravel bed perched high upon the mountainside, the remains of an ancient river.

Immediately he made camp and fell to work. With pick and shovel he worked away as he had done on so many previous occasions. The prospector is a hopeful creature, but he cannot always be at white heat. He learns not to "kid" himself and in the midst of hope is always braced to meet disappointment. So it was with Dr. Lloyd as he swung his pick on that lonely mountainside on that summer day. To him it was just routine, but as he raised his pick to strike another blow a familiar gleam caught his eye and the instrument fell clattering from his hand. Down on his hands and knees he went. The gleam was from a nugget which his pick had crushed. He dared not put his hand on it for fear it might vanish as a cruel illusion. He took his faded bandana handkerchief from his neck and mopped his brow and rubbed his eyes. He looked again. The nugget was still there, mocking him with its gleam of yellow light. Suddenly he seized it and ran wildly from his digging into the open spaces of the mountainside. There he unburdened his pent up spirit with every whoop and halloo that his boyhood mountain life had taught him.

With music in his heart and a song on his lips he now turned to the work of staking out his claims. For the first time in many a year he felt the same ebullition of spirit which he had experienced that day when he was able to go in swimming without taking his gun along, and much for the same reason. He was tasting freedom again—not freedom from bodily harm, but freedom from poverty, the world's cruelest slave driver.

Omitting tiresome details, it is sufficient to say that Dr. Lloyd laid out his mine and worked it systematically and, while it did not prove to be one of the world's outstanding gold finds, it did net the doctor a profit of several hundred thousand dollars. It did more than that. It established him as a successful seeker of treasure and placed him in demand as a partner in other enterprises.

[*NOTE: Lloyd was loosely playing on the legend of the Lost Lemon Mine of Alberta, Canada in the discussion of his purported Alaskan adventures, likely choosing it because a highly publicized expedition had gone out in search of the mine in 1929. Frank Lemon was quite dead by 1891 when Lloyd claimed to have prospected with him. Furthermore, Lloyd, as Joseph Durham, was living in Cincinnati in 1891, serving in the National Guard and had begun his medicine show. He had yet to visit Idaho and there were obviously no Shoshone Indians in Alaska for their guides to fight.*]

It was soon after this, in 1890 to be exact, that the noted prospector and adventurer, Frank Lemman, returned from Alaska. He came to Lloyd with a glowing tale of a rich gold find which he had made in that well-nigh impenetrable country. They set to work to organize an expedition to make the drive for this new source of riches. This was many years before the days of the "Alaska Gold Rush" and to most people it was a very uninviting prospect, but to Dr. Lloyd its very difficulties were its greatest attraction.

To Alaska With Lemman

After more than a year devoted to the task of getting their affairs in shape and assembling and supplies, they chartered a small steamer and set out. They went by way of the "inside passage" which at that time was not so well known and required careful navigating. At the end of some weeks they had worked their way about as far north as they thought the ice would permit. There they landed and built a camp at the mouth of a river.

As guides, they had brought along two Indians of the Sioux tribe of Idaho. These were Dick Two Elks and Plenty Wounds, young chiefs of their tribe, skilled woodsmen and bold hunters. They had often, with other members of their tribe, wandered far up into Canada and Alaska and they were familiar with both the country and its native Indian inhabitants, the Algonquins. From this circumstance, however, arose considerable trouble.

Two Elks was a very vain Indian who boasted much of the fact that he had slain three Algonquins at some time in the past, which he proved by carrying three Algonquin scalps at his belt, along with many others. This display of his prowess stimulated his fellow guide, Plenty Wounds, to a determination to bring his own record to par. In this attempt he waylaid a band of Shoshones, killed one, wounded another, and aroused such a furor that the party had to make a rapid march out of the Shoshone country to save themselves.

On the march, Lemman fell sick of fever and died. The little party dug a shallow grave in the frozen ground and buried him. Then they pushed on over the Chillcutt Pass and made headquarters on the Copper River. However, without the guidance of Lemman they found it impossible to locate the golden riches which he had discovered, and after a season of hardship and near-famine they returned to civilization…

Dr. Lloyd's success in the mining game had by no means been accidental. Back of it was the foundation of his scientific knowledge. He had started out as a chemist, which had enabled him to do assay work, but the broader aspects of the work drew him on and he took up the serious and technical study of mining engineering. From this it was but a step to the broader and more comprehensive study of geology.

STUDY OF GEOLOGY - FROM GOLD TO OIL

[NOTE: Lloyd, as Joseph Durham, was in Adrian, Michigan at this time, practicing "medicine." From there, in 1906, he headed on to Fort Worth to sell real estate in Arlington Heights and assume the name A. D. Lloyd. Posing as a petroleum geologist was still a decade away for him.]

His study of geology was such as had seldom been pursued by any student. With a foundation of academic scholarship and a thorough grounding in scientific method, he coupled a wealth of observation and practical experience seldom possessed by the same individual.

In 1904, he left the gold-bearing West and with it the mining game, and returned to Ohio. There he began his work in petroleum geology. Ohio, Indiana and Illinois were the scenes of his first efforts in the new work which he had laid out for himself, but his first discovery of an oil field was in the state of Kansas.

After this he ranged across the territory of Oklahoma, Texas, Arkansas, Louisiana and Old Mexico, in all of which regions he was responsible for the discovery of new pools of oil.

A Record For New Fields

In 1921, the writer met Dr. Lloyd in El Dorado, Arkansas, where he was doing geological work. He was very active there and much was said about his work. However, there is always an air of mystery about geology and one is inclined to wonder as to its real value. Is it really possible for geology to give advance information of the presence of an oil field?

For this writer this question was answered in a peculiar way, and thereby hangs a story.

Meeting Dr. Lloyd

In the early spring of 1921, the writer, with a group of friends, was seated in the lobby of the Garrett Hotel in El Dorado, Arkansas. Looking up, he saw a huge bulk of a man entering the doorway. So impressive was his appearance that one could not help asking his name.

"That's Doc Lloyd, the geologist," replied Tom Lewis, one of our party who seemed to have some previous acquaintance with him.

"Rock-hound, eh? Well, he looks as big as Taft and about as hearty."

"Wait till you see him eat, and for gosh sakes don't call him a 'rock-hound!' That big boy knows his stuff." This was said with a tone of admiration, but further remarks were cut short by the near approach of the Doctor himself. Just then he slapped Lewis on the shoulder.

"Well, young man, where have you been since the last greeting? Let's see, that was Ranger, wasn't it?" Then his eye took in our group and he said:

"I'm just from the woods; had a hard day; bring your friends along and have supper with me."

Of course we demurred, but it was no use.

"I'm lonesome and want your company," he said.

Now, if there is a group anywhere who can always be depended upon to appreciate an invitation to eat, it is to be found among the followers of an oil boom. Especially this is true at the beginning of operations, before the boys have had an opportunity to cash in on the play. Then it is that they are husbanding their resources and getting by on doughnuts and coffee and an occasional hot dog. Needless to say, we did not require much urging.

Six of us sat down around a large table, and Dr. Lloyd took charge of affairs. It was evident that he was in the habit of taking charge, wherever he might be.

"Boys, you may order what you wish, of course, but as a physician I'm going to prescribe a T-bone steak from the biggest cow in Texas. And, waiter, bring us about a half-gallon of french-fries and a basket of tomatoes, and you had as well put the coffee pot right here by my side."

Well, it was a glorious meal, eaten by a bunch fully able to enjoy it, but the meal wasn't the half of it. The conversation was the thing, and Dr. Lloyd took complete charge of that. Conversation belongs to him who has something to say, and on that score he was entitled to the floor.

While we were waiting for the food, he started up. It was a marvel of energy that we saw in action. His muddy boots and travel-stained field clothes gave evidence of the fifteen-mile hike that he had made through tangled woods and he should have been very tired, but he seemed to be just getting well started for the day. He regaled us with adventures in Alaska, Mexico and South America. He led us to various oil fields. His active mind seemed to forget nothing. He explained structures, described formations and expounded his theory of the origin of petroleum.

At last the delightful meal came to a close. As we were saying goodnight, Dr. Lloyd paused a moment for a final word.

"Boys, I've been making a survey of the country out east of here. Don't neglect that territory. It looks good. I'll leave you now. I have to work all night making a map of the country I've been over today." Energy? Well, it takes a lot of beefsteak to support that sort of driving power.

NEGLECTING GOOD ADVICE

[NOTE: Development of the fields east of El Dorado actually began in 1921. Lloyd's prescient advice was of events that were already in progress.]

"Don't neglect the territory out east." Did we heed his advice? Did anybody ever have sense enough to recognize and follow good advice when it was free? Certainly not. We could think of nothing but to follow the crowd in its mad scramble west of town.

We stood by and watched others bring in the East field in 1922, the Rainbow field in 1927 and Urbana in 1930, all in the very territory which Dr. Lloyd told us about in 1921. Instead of profiting by his friendly advice, we saw a new crop of millionaires spring up on ground that should have been ours. We saw a dentist, who had not received any free advice or beefsteak either, take a gambler's chance, mortgage his tooth-pulling outfit to the limit, and make a half million dollars out of a twenty-acre lease in the midst of the Rainbow field—and we hadn't enough sense to take a little free advice from a man who knew.

It was in the spring of 1930, after the Urbana field had come in, that this writer visited the office of one of the larger oil companies at El Dorado, in search of information on that region. The young geologist in charge was very courteous and helpful. He furnished logs of wells and other useful data. Then he took from his files a much-wrinkled map and spread it on the table.

On the map was located the East field, the Rainbow field and the Urbana field. The map was dated 1921 and was signed by Dr. A. D. Lloyd, Geologist, El Dorado, Ark. Patently it was the map drawn by him the night of our big feed. He had been a year ahead of the East field, six years ahead of Rainbow and nine years ahead of Urbana. As I stared at the map, it seemed a thing alive and able to speak, and I halfway expected it to rebuke me for my lack of faith. I felt that the dapper young geologist of the big company was reading my thoughts. Finally, I managed to give voice to a question.

"How does he do it?" I asked. The young geologist shook his head.

"Search me," he remarked. "He made this map before there was ever a well drilled in that region. Now we have the logs of several hundred wells to guide us and can make little improvement on his work. I don't know how he does it."

"Well," said I, "I'll tell you how he does it. He gets out there in the field where the actual conditions are and sees for himself. Then he interprets them according to laws of science which he has mastered and knows can't go wrong.

He does this while you fellows are riding a swivel chair and following the latest scientific fad laid down by someone else. While you are putting in your time condemning structures and crying 'impossible' he will be bringing in new oil fields. How do I know? I'll tell you how I know. I ate beefsteak with him and heard him tell about all this the night before the map was made."

This circumstance set me to thinking. At last I knew there was something valuable to this science of geology. I had seen Dr. Lloyd's name mentioned several times of late, in connection with Lloyd Oil Corporation. I decided to come to Fort Worth and find out what it was all about.

I have for a long time wanted to write a story of oil in a style that could be understood by the layman. I conceived the idea that this could best be done by writing the history of one particular oil company, using it as a type. Considering my experience with Dr. Lloyd, I decided to use Lloyd Oil Corporation to represent the type.

Dr. Lloyd on the Gulf Coast

Back to the geological work of Dr. Lloyd, we find that after his work in Kansas he decided to come south. Oil had already been discovered on the Gulf coast and he turned his attention to that region. Beginning near New Orleans, he made a careful survey of the coast country bordering Louisiana, Texas and Mexico as far around as Yucatan. It required two years for this work, the dates being 1908 and 1909. The latter year he spent in Mexico.

During those two years he mapped many of the coastal salt domes which have since become prolific producers of oil. In Mexico he was no less successful. His work there so impressed the Mexican officials that he was given valuable mining concessions, which he proceeded to develop. Afterwards he was appointed to make a geological survey of the Republic.

When Dr. Lloyd went to Mexico in 1909, disaffection and rebellion was just rising to unhorse the "Iron Man" of Mexico, Gen. Porfirio Diaz, who had been president of the country for thirty years. Under him everything had been tranquil, and Dr. Lloyd opened his mines without difficulty in the Iguana Hills, at the head of a green valley in the Sierra Lampasas Mountains. Being a green valley and blessed with water, it was a favorite rendezvous for bandits of that region. Soon it also began to receive the backwash from the revolutionary movement which was sweeping the country.

Joining the Revolution

It is difficult to tabulate all the influences which help to shape our course of action. Frequent raids, lax enforcement of law, and consequent loss of property were sufficient to goad Dr. Lloyd into some measure of retaliation, or, at least, of self-defense. Accordingly, he threw in his fortunes with the revolutionary forces. He had been in the mining business with a nephew of Madero and it was an easy matter to establish friendly relations with that rising leader who was soon destined to become the president and dictator of Mexico. Through varying fortunes he was allied, in turn, with Huerta and Carranza.

It was as a member of President Carranza's staff that Dr. Lloyd attained his greatest influence in Mexico. Carranza appointed him to make a geological survey of the petroleum resources of the Republic. In carrying out this work he mapped many of the structures which have since become rich producing fields in that country.

Since his first visit to Mexico, in 1909, Dr. Lloyd has spent about one-third of his time in that country. Fate seems to have cast him to play a part always on the frontier, where the action is most intense and the conflict most determined. As a pioneer he went to Idaho, to Alaska, to Mexico and to South America. Always he was engaged in exploration, in search of the riches which had been stored up for man's use. His life has been a struggle to harness nature's forces and make them to contribute to the enjoyment of the human race. In this he has generally succeeded.

One may pause to wonder over the ancient debate between heredity and environment. Was this passion for discovery and conquest a result of some primeval urge handed down to him from those hardy ancestors who at an early date made their way into the Appalachian wilds and fought and won their battle with nature; or, was it merely a logical outgrowth of his own youth and of the hard circumstances which had compelled him as a boy of nine to take up arms in defense of himself and his family in the feudist quarrels of "bloody Breathitt"?

Mexican Adventures

The Lloyd mining camp in the Lampasas Mountains was organized along the usual lines. There was the array of huts as quarters for the peons, the casa grande or "big house" for the owner, the commissary from which the laborers received their supplies, a convenient church, and a still more convenient saloon with gambling house attached. Without these last three institutions it was impossible to hold a force of laborers together.

Pay day came once a month but rations were handed out oftener. Saturday night was the big occasion for sport. Then would drinking, gambling and fighting rule the place. As is well known, the Mexican prefers to fight with a knife. It is a cheaper weapon, is more easily concealed, does its work noiselessly and seldom misses.

Along toward midnight when the mescal had gotten in its work, it would be easy for some Mexican to fancy that another was cheating at the game, or to remember some fancied injury directed at his dignity on some other occasion. Then the knives would flash, blood would flow, the crowd would take sides and pandemonium would reign for a time. In fact, it usually held sway until someone recovered his wits sufficiently to send for Dr. Lloyd, who was in his headquarters at the "big house."

A peculiar law in Mexico prohibits anyone except a physician from touching a wounded man, even though he may be about to bleed to death. Consequently, when a man was wounded in one of these brawls he was permitted to lie where he had fallen, lest the one touching him should be charged with causing his death. Doctor Lloyd would arrive on the scene and, through his authority as a Doctor of Medicine, would lay the wounded out on cots, kept for that purpose, would administer first aid to those needing it, and would stow the dead in a convenient back room under an improvised shroud, there to wait the coming of day.

These duties having been performed, he then turned his attention to the work of soothing the anger and quieting the fears of the survivors of the fray. For this there was one unfailing formula. He invited all those present to come up and have a drink on the house. This was accompanied by cries of "Viva la Doctor! Viva la Boss!" after which they understood that the entertainment of the evening was at a close, and made their departure.

On these occasions there were always present some of the bandits who were making their headquarters in the valley, but they never offered to molest anyone. In fact they had never made any demands on the mine or its owner, further than to ask for credit at the commissary and saloon from time to time, and these bills had always been paid.

An Escape From Kidnappers

There came a time, however, when things were different. Returning from Mexico City on one occasion, the Doctor was met at the railroad station some eighty miles from his mine, by his mine superintendent, Felipe Naranjo, who excitedly informed him that the bandits were waiting for him on the road with the intention of holding him for ransom. Felipe begged him not to go out that day but to wait at the station until the bandits had dispersed.

The Doctor thought the matter over and decided that he would gain time by facing the situation at once. Climbing on the buckboard, he set out with all possible speed toward the mine. At the gate of a large ranch through which he was to pass, he found the outlaws waiting. They were well known to him, being led by Juan Cruz, who had frequently received credit from him.

Without waiting for them to make a demand, the Doctor pulled out a bottle of choice liquor and passed it to Cruz. That worthy, unable to resist such hospitality, took a generous drink and passed the bottle to the next in command. When it had gone the rounds, the Doctor took the initiative in discussing the business at hand.

"I understand that you boys have planned to hold me for ransom."

"Yes," they admitted, "they had planned a matter of that kind, not hostilely but as a mere means of livelihood."

"Now listen to this," said Dr. Lloyd, "have I not extended you credit when you needed food? Have I not let you have liquor and ammunition and feed for your horses? What more could you want? It seems to me that you had better save me for emergencies, for hard times, and go rob somebody else now. Have you ever robbed the mine of Don Epermenia?"

"Don Epermenia? No, we have not." There was some rapid consultation. Then the leader turned to him.

"Señor, your suggestion is a good one and we thank you. We guarantee you safety and shall save you for a more critical time."

With that they rode away. Next day came news that the mining camp of Don Epermenia, ten miles across the mountains, had been most thoroughly looted by Cruz and his band.

GUEST OF THE MEXICAN GOVERNMENT

Dr. Lloyd still retains his interest in mines in different parts of Mexico. Also, he has received concessions on properties for exploring for oil and has contributed much to the development of that country. He has a host of personal friends there and has a sincere admiration for the people and their culture.

In January 1929, Dr. Lloyd and the officials of the Lloyd Oil Corporation, of which he is president, received a special invitation to attend the inauguration ceremonies of

President Portes Gil. They went to Mexico City for the purpose, and received every courtesy at the hands of the government. Not only were their physical comforts supplied, but a General of the Mexican Army was assigned as their daily escort. When the days of celebration were over and they were about to depart, they were presented with valuable oil concessions for future development.

It requires a man of more than ordinary personality to impress himself on a foreign people and win the approval of those whose outlook on life must necessarily vary greatly from his own.

If we were called upon to give reasons for Dr. Lloyd's success in this respect, it would be our judgment that the dynamic force of his personality, existing as a primitive and fundamental element of his nature, rises above the limitations of race and language and builds and enduring foundation of confidence and friendship—that his essential integrity and his good intentions are sensed by these people though their ideals and philosophy of life be ever so alien to his own.

An entire treatise could be devoted to the discussion of the influences which are able to take a youth, stunted by the hard, narrow and embittering isolation of the mountain feud country, and expand him into a genial, tolerant and attractive citizen of the world, but such discussion is beyond the scope of this sketch.

MANY-SIDED CHARACTER

Whatever the origin of his tendencies, the fact remains that life's reactions have developed in him very profound characteristics and have endowed him with a rich personality.

Of large physique, he is naturally impressive in appearance, but nature endowed him with a mental quickness and grasp and a power of will which makes him easily dominate any situation which may arise. Because of this faculty he is as much at home in his human relations, in his conflicts and dealings with men, as in his scientific pursuits.

There is about him a spirit of camaraderie and good fellowship which easily disarms opposition and gives him entre into the minds and hearts of all. It may be analyzed as a spirit of tolerance and benevolence, a feeling of sympathy for the weaknesses and griefs and hardships which others endure, a feeling which arises from the grinding toil and the heartaches which he himself has experienced in his onward march toward achievement.

There is nothing petty or little or narrow about Dr. A. D. Lloyd. He early caught a vision of the bigness of life and he moulded himself to fit it. Generosity of thought and deed have always marked him. In his attitude he has something of the graciousness of the old Southern planter, the breeziness of the Western cattleman and the generosity of the mining baron. In any company he would be denominated "a good sport."

Here again old Kentucky enters in. In his lighter moments, no doubt, "strong liquor, beautiful women and fast horses" have played their part.

There was the fine racing stable at Hailey, Idaho. What became of it? He gave it to a friend when he decided to forsake gold mining and take up oil. Why not? It had served its purpose as a means of relaxation and he could not go dragging a bunch of racers around among the oil fields of the continent. Besides, what was a mere bauble like that to a man who always felt that he had the power to go out and make a fortune at will?

What became of the racing stable in Cincinnati? He donated that one to his partner. Why not, again? To his partner it was a sole means of livelihood, while to him it was merely a diversion which probably had served its purpose for the moment.

Then, there was the birthday party at Hailey. It was in that ancient period, now sometimes alluded to as "the good old days," before prohibition. The Doctor was having a birthday. Unwisely he made mention of the fact to some crony. The news spread rapidly. It seems there was a custom of which the Doctor was quite ignorant. In any "wide-open" town everything serves as an excuse for taking one more drink. Especially is a birthday regarded as a most appropriate time to celebrate, and always at the expense of the one having the birthday.

The Doctor's reputation for generosity made him a very attractive victim on such an occasion. Soon his favorite bar was filled with a host of admiring and clamorous friends. "What will they have?" "Well, they would start with a beer." The Doctor looked the crowd over. He looked down the street and saw several hundred more approaching in hilarious holiday spirit.

"All right, Jake, this is on me. Roll out ten barrels of your best brand, knock the heads out and hang a bunch of dippers on every barrel. We don't want anything to slow this party up."

Oil Fields To His Credit

We have mentioned that Dr. Lloyd's first oil field discovery was in Kansas. In a general way we have taken notice of his work along the Gulf coast and Mexico. To be a little more specific, we now name some of the well-known fields which resulted from his studies. Chief among these are: The original Brown County field, the first Texas field to be opened up away from the coast; the Mercury field in McCullough County; the great Panhandle field of North Texas; the Breckenridge field; the Necessity Anticline; the Sipe Springs field; the Fry Pool; the Pilot Point and Wheelock fields, in Cooke County; the Coleman County field; and, latest of all, the great East Texas field, which was uncovered by Dad Joiner, following the survey and map made by Dr. Lloyd more than three years before the well was brought in.

In addition to these Texas fields, the fields at Cement, Newkirk and Earlsboro, Oklahoma, were the result of Dr. Lloyd's geology, all of them brought in by Dad Joiner. The Cement field is particularly noteworthy as being the first field brought in in the red bed district of that state, in the face of the general belief among oil men and geologists that there was no oil in the red beds.

Including the fields in Arkansas and those along the Gulf Coast and in Mexico already alluded to, the total number of fields designated and mapped by him, in advance of drilling, will number more than thirty. Considering the fact that the average geologist thinks himself well-paid if he succeeds in locating one paying field in a lifetime, the performances of Dr. Lloyd are truly remarkable.

When asked for an explanation of his amazing success in locating new fields, Dr. Lloyd was silent for awhile, apparently absorbed in thought. At length, he replied:

"I suppose my success in finding new oil fields may be attributed mainly to my theory of the origin of petroleum. If a man wishes to find an object, it is easy to see that it would give him a great advantage if he only knew the source or origin of the object which he is seeking. He certainly would have a better idea where to look.

"Early in my work as a geologist I became dissatisfied with the prevalent idea that petroleum was derived from vegetable or animal life. These sources seemed pitifully inadequate and the conditions under which we are finding oil seemed all out of joint with the conditions under which it would have been formed from organic sources. This put me to thinking and observing and in the course of time I gathered data which convinced me that it had an entirely different origin. This material I published in 1918, since which time my theory has been accepted by many people. With this theory of origin as a working basis I have been able to accomplish my work in the oil fields."

The writer has listened with much interest to the unfolding of Dr. Lloyd's theory of the origin of petroleum and then secured a copy of his original paper on the subject, which is printed in the following chapter.

Dr. Lloyd's Theory of Origin of Oil

Anyone making a study of Dr. Lloyd's operations in the field of petroleum will be impressed with the idea that there is nothing haphazard about it. We will soon see that the discovery of thirteen proven fields in Texas, and probably as many more scattered over other states, together with his discoveries in Mexico, was not a freak accident. Years of scientific research, years more of observation of facts as he met them in the field, and years of contemplation and logical reasoning brought him to certain scientific conclusions which have been the basis of his decisions with reference to possible oil structures.

Among the principles which have guided him in this work, probably none has been more important than his theory of the origin of oil. It was first published in 1918.

[NOTE: No publication of the following, Lloyd's treatise on the origins of oil, has been located in the oil publications of the day or in the Library of Congress.]

THE ORIGIN OF PETROLEUM
by
A. D. LLOYD

At the time when I published my treatise on the saline domes of Texas and Louisiana, the fields then producing were obtaining their oil from the limestone cap rock secondary from the salt core, all at about the same depth, being approximately 1,150 feet. If they went through this secondary concretionary limestone, they cut 1,000 feet of shales and conglomerates before reaching the salt mass. Salt domes so developed included Spindletop, Sour Lake, Saratoga, Batson, Humble and Markham in Texas; and Vivian, Jennings and Anse La Butte in Louisiana. The domes which were more uplifted, such as Damon's Mound, Dawn Hill, Hackberry Island, Barber's Hill, Blue Ridge, Pierce Junction and Big Hill, in Texas; and, Five Islands, New Iberia, Rayburn Saline, Drake Saline, Winnfield Marble Quarry and others, in Louisiana, had been drilled and the salt core encountered at a depth varying from 100 to 600 feet. Such domes did not produce any oil.

The general conclusion of geological engineers investigating saline dome oil deposits was that the salt core was a concretionary mass. Some of these masses had been drilled to a depth of two or three thousand feet, but none of the salines of greater uplift had produced any oil, although six wells had been drilled on Damon's Mound. However, I obtained cores with igneous intrusives and substantiated the presence of porphyry granite found at a depth of ninety feet in the sulphur saline at Sulphur, Louisiana, where this saline is being mined for its sulphur content. The superintendent of the mine gave me a sample of this igneous rock, from which I advanced the theory that the secondary crystalline limestone known as the cap rock, from which oil was being obtained in the saline domes then producing, had been formed at such a depth as the downward-percolating surface waters had reached before becoming supersaturated with calcarious matter, at which point they came in contact with the gases from around the deeply buried saline mass, the effect of this contact being to precipitate the lime from the water thus so heavily saturated, and in this manner to form the cap rock, the very crystals of which were saturated with oil.

In this treatise, I advocated that test wells should be drilled around the circumference of the saline of greater uplift, believing that the same secondary limestone, deposited by a similar method, would be found at a greater depth, where the precipitating process had occurred, and that such uplifted salines, which had been dry in previous tests, would in this way be made oil producers. Following this theory, a friend of mine

(George Hammon) backed off from Damon's Mound and drilled a five thousand barrel well which stood for years. Likewise, a friend of mine backed away from the producing Humble field and at a depth of 3,000 feet got a ten thousand barrel well. Going out with him the night the well came in, I asked him why he had drilled this well off the old cap rock district. His answer was that he had read one of my damned books and he thought that, if it held good on the dry ones, it ought to hold good on the wet ones.

In my treatise on the Gulf Coast and Louisiana, I set forth the possibility of using the seismograph in searching for the oil deposits around the saline domes. Today the major companies are using the seismograph for saline domes exclusively. Since the publication of my treatise, many new saline domes have been developed into oil producers, by following the theory which I laid down and the principles of my original treatise have been accepted by the petroleum engineers of today.

The origin of oil was explained by three different theories set forth by various petroleum engineers and geologists. One having the largest following was the theory of vegetable origin. The second was the animal origin, accounted for by a great cataclysm which had destroyed and entombed millions of fish and animals, and that wherever fish and animals had congregated, the oil was created. The third theory was that oil originated from sea algae and diatoms.

I set up the hypothesis that oil was of volcanic origin. I established that petroleum oil contained only two elements, carbon and hydrogen, while vegetable and animal oils contained three elements, carbon and hydrogen and oxygen.

The diatomic theory of origin had been built around the fact that while the diatom was alive the algae would show the presence of petroleum by refraction of light or radiolarity. The great difficulty with this theory was that it was impossible to actually extract any petroleum from these animals by chemical process, and furthermore that when the algae was dead it even lost all refractory evidence of the presence of petroleum.

The advocates of the theory of vegetable origin contended that the vast quantities of floral debris accumulating in the marshes afterwards received a covering of sludge and was ultimately transformed into petroleum oil. On the other hand, it was my contention that no chemical process could be demonstrated in the laboratory which would remove oxygen from the vegetable oil and transform it into hydrocarbon rich petroleum oil. Further, I offered as evidence that when wood is buried it is preserved indefinitely as wood fiber, and that it never turns into peat, lignite or oil, although peat and lignite both contain the three elements, hydrogen, oxygen and carbon. In the Humble oil field, pecan logs and cypress knees are cut at a depth of 900 feet, and pecan, hickory and oak at about 400 feet. In the Oligocene trough of the Mississippian Embayment, a test well cut forty feet of timber in a perfect state of preservation at a depth of three thousand feet. The remains of trees found in coal mines are never coal. They are always flint or sandstone, showing that wood fiber and its oily contents never went to coal or oil.

A further substantiation of my theory of volcanic origin was the fact that in large oil fields like Humble, Goose Creek, Saratoga and Sour Lake, when oil was lost and flowed off down the creek, it was totally oxidized and destroyed by the time it had flowed as far as five miles, leaving no trace of oil on the drift wood or rocks or muds of the stream. This indicated that oxygen destroys petroleum, and that it never could have been created from any oil that contained oxygen. Oil can be extracted from any vegetable in some amount, usually varying from five to forty per cent, while the vegetation is alive, but when it dies and disintegrates, it is totally destroyed and leaves no deposit or trace of oil behind, nor does it show the presence of oil at any stage of its decay and disintegration.

When oil is put in a storage tank, say one of 55,000 barrels, it will begin to take on oxygen at the top and will be oxidized and destroyed, forming a crust as much as a foot thick and strong enough to bear the weight of an average man. As to the animal theory, there is no evidence in any of the fields of catastrophes that could have entombed fish or animals to make the millions of barrels of oil produced by those fields.

The primeval rock, or original crust of the earth, was granite, and that rock contains all the elements known to chemistry. It contains carbon in the form of graphite. From granite blown out of an active volcano, I myself have extracted as high as four per cent of oil. From basalt, the hardest of igneous rocks, I have taken globules of oil as large as a cubic centimeter. I have seen oil seepage in mines of porphyry granite. I have seen oil seepages in the craters of extinct volcanoes. The diamond, which is pure carbon, occurs in volcanic muds and ashes. These facts show the presence of the one element, carbon, in the neucleritic crust of the earth. The felspars of partly disintegrated granite will take on oxygen with great avidity. Steam or vaporized water illustrates the formation of all volcanic gases. It is my view that the water, which is composed of two elements, hydrogen and oxygen, finds its way along faults or fractures in the earth's crust to the submerged molten, igneous masses, where the oxygen of the water goes into combination with the basic felspars of the granite and liberates its hydrogen, which recombines with the carbon in the granite to form petroleum oil, which is composed of these two chemical elements, carbon and hydrogen; and, that, through the ascensive processes of volcanic action, petroleum in the gaseous form moves out laterally from the circumference of the igneous core; that in this movement it encounters lower temperatures and is ultimately condensed to form petroleum oil.

I contend that the greatest source of oil is the laccolith, an igneous disturbance prevented from acquiring sufficient ascensive force to become an eruptive volcano. In the oil fields of Kansas, Oklahoma and Texas, evidences of such granite ridges are found. Crossing the state of Kansas, there is a laccolith band twenty miles wide, where metamorphic granite is encountered at depths varying from 900 to 2,000 feet. This constitutes a very large laccolithic intrusion into the sedimentaries of Kansas, which are from 15,000 to 25,000 feet thick, and the oil is found on both sides of this gran-

ite ridge. Likewise, there is a granite ridge extending
across Oklahoma and the Texas Panhandle. It appears
in Oklahoma twice, once in the Arbuckle Mountains
and once in the Wichita range. Laterally, surrounding
these laccoliths, the great oil fields of Oklahoma are
found. In Texas, they cut this granite at 1,200 to 1,600
feet across the entire Amarillo oil producing district.
In the oil field country of Luling and Thrall, and in
the southeast corner of Travis county, along the Colorado river, all centering around San Antonio, there are
exposed laccoliths. Also, laccoliths are found exposed
in the Guadalupe river. As long as volcanic processes
occur, the elements of earth's oil will continue to be
created and accumulated.

COMMENT

To one who has not spent much time in oil fields or in a
study of petroleum geology, it will probably be difficult to
explain just how revolutionary the above theory accounting for the origin of petroleum really is. To say that it is the
exact opposite of all other theories comes near being the
truth. If you want to give the average oil man a fright, just
tell him that he is drilling near a granite intrusion. He has
been brought up on the theory that if there is any volcanic
material near at hand, there could not possibly be any oil
because the volcanic matter would long ago have burned
the oil up. Now he is being told that that same despised
and feared volcanic material is the very source of the riches
which he is desirous of uncovering.

In scientific progress, the world has made its way by first
observing details and then slowly, step by step, piecing

them together to form the whole picture. This is equally true with respect to solving the puzzle of the origin of oil. The two great divisions of rock are the volcanic and the sedimentary. The drilling experience of the oil industry teaches that the underground bodies of oil which constitute our oil fields are practically all found in sedimentary or water-deposited rocks. We must therefore differentiate. The sedimentary rocks are the reservoir or store house of oil because they are porous, because their more or less regular layers act in the capacity of pipelines to conduct the petroleum gas away from the place of volcanic origin, and because the sedimentary rocks are bent into anticlines and other containing structures capable of accumulating the oil in large quantities. It is this storehouse that we have been best acquainted with and for which we are constantly seeking. The place of manufacture, the chemical plant where oil is made, is the volcanic rock which we have been shunning. The impervious nature of most volcanic rock would prevent it from becoming a storage place for oil, but being the originating place for it, the storage will doubtless occur in the sedimentary rocks close by. Hence the significance of finding granite and other volcanic material so near many of our oil fields. Hence, also, the significance of the theory of the volcanic origin of oil. It tells where to look for its reservoirs...

DR. LLOYD'S DISCOVERY OF FAMOUS EAST TEXAS FIELD

For some unaccountable reason quite a controversy has been raised by a few individuals as to credit for pioneering the East Texas Oil Field. One geologist of much book learning and considerable ability as a writer for the papers has constantly harped on the statement that the discovery was an accident, pure and simple. What his motive is we do not know. He may be one of the great host of geologists who constantly threw cold water on "Dad" Joiner's attempts to raise money for his project with the statement that "there just couldn't be any oil in East Texas." Dad Joiner found a lot of that kind when he was struggling with his wildcat, but now they "knew it was there all the time." It might be that such reports could be started through envy. A man who had worked all his life as a "big time" geologist without discovering a single field might feel that way about a man who had been "lucky" enough to "accidentally" discover thirty-one fields during his lifetime.

However the controversy may have arisen, we give here the actual facts as they occurred and as they can be easily verified and leave it to our readers to decide how "accidental" the discovery was.

During the course of his eventful career as a wildcatter, C. M. "Dad" Joiner had brought in three new wildcat fields on geology furnished him by Dr. A. D. Lloyd. The first of these was at Cement, Okla., on a structure in the red bed area, where the geologists had been at great pains to declare there could be no production. The second was at the Earlsboro extension of the great Seminole field also in Oklahoma. The third was the Newkirk field.

Now, it happened in the spring of 1927 that "Dad" Joiner was cruising around in East Texas, looking the country over. Certain parties there, knowing that he was an enterprising wildcatter, approached him with a request that he drill them a wildcat well. He liked the general appearance of the country but told them that he would not drill except on territory approved by Dr. A. D. Lloyd. He then set out to find Dr. Lloyd. He came to Fort Worth in search of him. At that time Dr. Lloyd had not yet organized the Lloyd Oil Corporation, and he was not maintaining an office in Fort Worth. In fact, he was spending his time in the Republic of Mexico. After some inquiry among mutual friends, someone said: "There isn't any telling where Dr. Lloyd is, except that he is out somewhere trailing down a new oil field, but it's a cinch that he will come back to Fort Worth, and when he does come back, he will show up in the lobby of the Texas Hotel. If I were you I'd just get me a seat and wait right there."

Patience Rewarded

Dad Joiner followed the suggestion. He just "took a seat and waited." It was nerve-breaking to a man of his active disposition, but he kept waiting. He waited there just sixty days, and then one fine day about the first of May, 1927, Dr. A. D. Lloyd loomed up in the door of the Texas Hotel. The Doctor was probably at a loss to understand the warmth of the greeting extended him by Dad Joiner, in spite of their long years of friendship. He could not know how welcome a man is after he has been awaited for sixty days in a hotel lobby.

"Well, Doc, you're going to East Texas with me," were the first words that greeted him.

"I can't do it, Dad. I have a man waiting here to take me to West Texas. Been waiting two weeks now," replied Dr. Lloyd.

"Shucks, what's a little old two-weeks wait? I've been right here waiting two months."

"All right, you win. I'll run over and spend a couple of days with you."

117

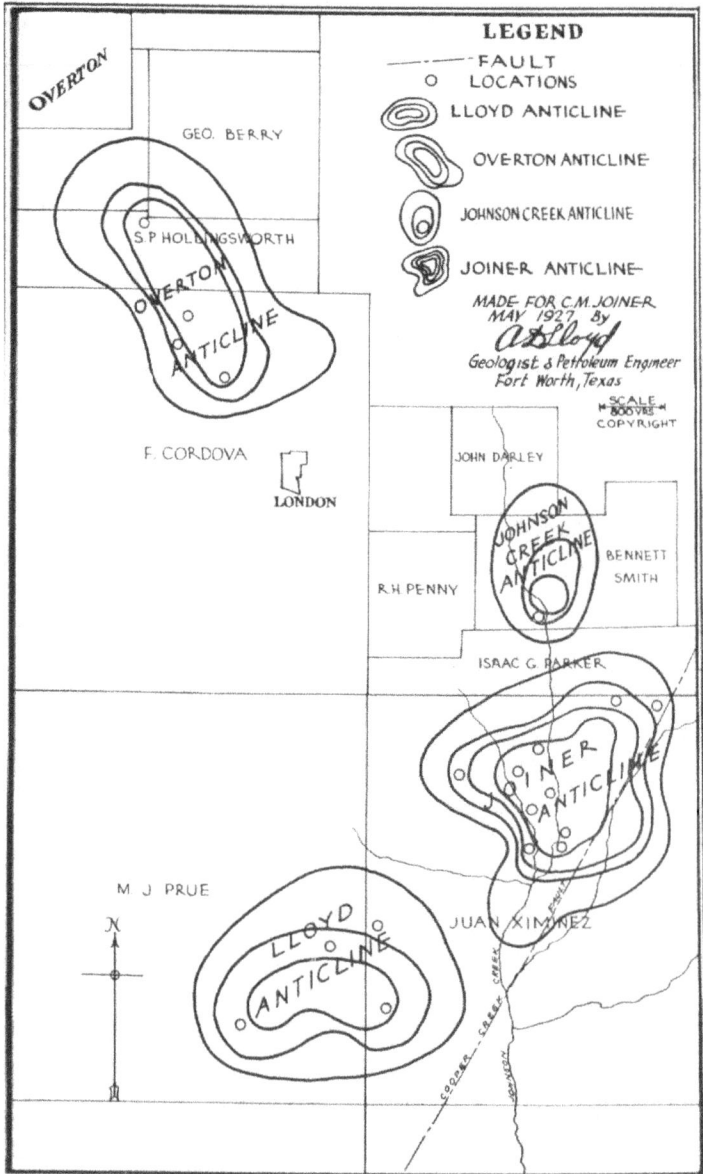

LEGEND

FAULT

○ LOCATIONS

LLOYD ANTICLINE

OVERTON ANTICLINE

JOHNSON CREEK ANTICLINE

JOINER ANTICLINE

MADE FOR C.M. JOINER
MAY 1927 By
A.D. Lloyd
Geologist & Petroleum Engineer
Fort Worth, Texas

SCALE
800 VRS
COPYRIGHT

Though similar to the map he produced for Joiner's East Texas promotion in 1927, this version provided by Lloyd for Cubage's 1931 book omits a salt dome and is slightly rearranged in Lloyd's favor.

118

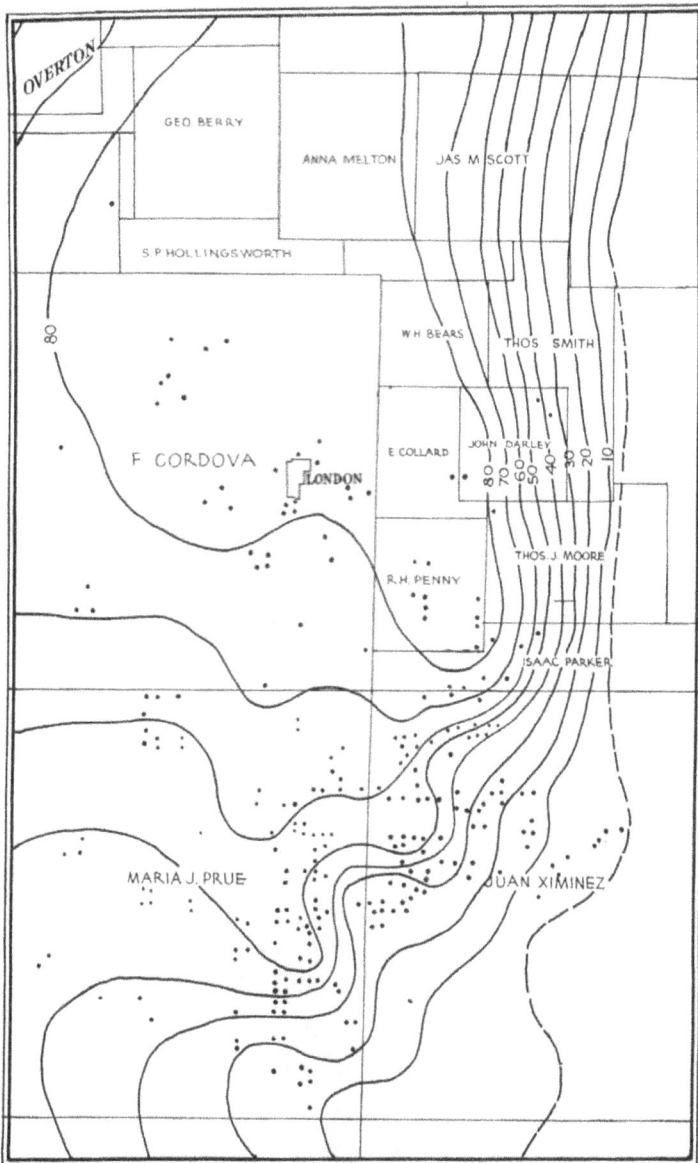

Second map in Cubage's 1931 book, demonstrating how the wells then drilled in the East Texas field correlated with Lloyd's locations on the facing map.

He went. He got interested in the conditions that he found and remained for weeks. Then he made a map of the territory which he considered to have oil-producing possibilities. He also wrote Dad Joiner a letter telling him what he thought of the structure which he had mapped for him. A copy of the map appears on a [previous] page. Compare it with the map facing it, which is a map of the same territory based on the wells that have since been drilled. You will notice that the highs indicated on the Lloyd map are located just where the highs were actually found by drilling, as shown on the second map.

Notice that Dr. Lloyd accurately defined the eastern and southeastern edge of the field with his locations.

The very small circles on the Lloyd map are drilling locations recommended by him and placed on his map in 1927. These locations have all been drilled since that time and are now producing oil wells...

Oil As An Investment

...It is only reasonable to think that an industry, such as the crude oil business, which produces a billion dollars worth of raw material each year over a ten year period would afford money-making opportunities to a vast number of people. It goes without saying that thousands of laborers have exacted a toll for their labors in the oil fields, that supply men, teaming contractors, construction men, pipe fitters, and a host of others have taken tribute from the business of producing petroleum, but there is an even greater number who have tasted its benefits. These are the ones who have furnished the capital necessary to keep operations going.

One of the problems connected with any growing business is that of providing capital fast enough to keep up with its expansion, and no business has had a faster growth than the oil business. To meet this demand for capital, two very different methods have been resorted to.

One method is to deal with the large financial powers and groups. They are usually approached through brokers or bankers, or through engineering houses. Under this method, there is much demand for commissions and cuts, and much watering of stocks and bonds to make room for all the interests that have to be taken care of.

The other favorite method is to appeal direct to the small investor without the intervention of so many middlemen. This method is more democratic than the first and would seem to be the natural way of keeping down monopoly and keeping the great natural resources of the country in the hands of the people.

It is this second method that Dr. A. D. Lloyd, president of Lloyd Oil Corporation, favors. His long and useful life has taught him a striking philosophy of wealth. The world long had an idea that poverty was the natural and inherent condition of man. "The poor you always have with you," to them was not only a statement of fact but a sort of blast from trumpets of doom which forever condemned men to that fate.

Dr. Lloyd's view of life is just the opposite. He believes that it is as natural and as reasonable that all men should be rich as it is that all or most should be poor.

Somewhat naturally, due to his long connection with the oil business, Dr. Lloyd thinks of wealth and its distribution in terms of storage tanks and pipelines. He conceives of all wealth as a liquid commodity stored up for the use and benefit of the nation and its people. Coal, copper, iron, lead, zinc, aluminum, sand, clay, soil, water power, petroleum, he sees flowing through great pipelines to their ultimate destination. These are the foundation of wealth and all people ought to be permitted to share in their exploitation and development. Any attempt to limit investment in the nation's raw material to a few favored groups of investors in Wall Street or in any other limited sphere he views with the greatest misgivings and fear for the safety and security of our future. To limit investments in natural products to such interests would throw our raw materials into the control of monopolists for ages to come.

Not only that, but such a policy would deprive the great mass of our people of one of the great privileges of life which the spirit of our government guarantees them. They have a right to "the pursuit of happiness by provision for their material well-being." He looks back at the course of business in this country and reflects upon the vital difference that would have come to the millions of our people if, in the beginning and continuing down to the present day, they had put their money into the companies which developed and controlled our coal, iron, lumber and other natural resources instead of turning these over to a few groups of exploiters to plunder and grow rich at the public expense.

If every possible investor had connected himself to that reservoir of natural wealth by a pipeline of stock investment, ever so small, it would have had marked effect upon national financial stability and upon the average prosperity of the ordinary family.

Steam, iron, coal, and lumber are already under firm control and no longer present attractive investment opportunities to the public, but oil, which is more important than any of them, still remains.

If skeptics would attempt to decry this, the facts are at hand to refute them and to lend proof to the statement. Even with the country passing through the greatest depression in its history, even with the oil business itself feeling that depression as no other business, even with oil selling at the lowest price in a quarter of a century, we find men and women making fortunes overnight.

The East Texas field is the newest of the great fortune-making oil strikes and it is natural to make mention of the events connected with it. This is all the more natural since it was due to Dr. Lloyd's geological work that this field was discovered.

Dad Joiner, after the usual run of wildcatters' hard luck, brought the field in. He was three years making the completed test, during which time he lost two wells and barely saved the third which was the discovery well. He will likely never know how much the venture cost him, but it was probably around $75,000. Reports carried in the daily papers at the time stated that he had sold out for just a little less than $1,000,000.

Ed Bateman, of Fort Worth, brought in the discovery well on the Crim lease, near Kilgore. The operation was quite extensive, costing between forty and fifty thousand dollars, but he sold out for $2,100,000, and he and his stockholders profited accordingly.

Moncrief and associates, in partnership with the Arkansas Natural Gas Co., brought in the discovery well near Longview. No figures are available as to the cost, but for their half interest it should not have been more than $25,000. They sold out their half interest for $3,500,000.

The desire for power is the strongest craving of the human soul. As the small boy wishes to grow up and be a policeman, even so does every man picture himself the future master of others. And most fortunate is that he usually envisions himself as using that power for good rather than evil—for, every man is at heart a philanthropist and yearns to do good to somebody.

It was Emerson who gave the definition for true philanthropy. Said he, "I would not give a man alms, but I would afford him the opportunity to earn." Opportunity! What a golden word! How every man's heart hungers for it! The opportunity to earn, to gain, to accomplish, to achieve. The chance to overcome the snares and pitfalls and failures which have eternally beset him.

And who are the great philanthropists? Who are the men who have opened up to their fellow men the greatest opportunities in life? Who, indeed, but that noble and consecrated band of men and women who, forswearing ease and comfort for themselves, have pledged heart and hand and brain to the cause of science? Forever delving into mysteries, they make bold strides to unlock nature's hidden vaults and drag forth her most precious secrets for the enrichment of all mankind. It is their work that has filled the world with new wealth and has made life full and rich and radiant everywhere. It is their work that has added efficiency to human activity, and has brought convenience and happiness to a world overburdened with backbreaking drudgery.

So, we say: All honor to the scientist. He has given us new metals, new elements, new tools, new machines, and new sources of power. He has enriched us beyond our most daring dreams and has made overproduction of the necessities of life our only calamity. Yes, let us sing a pean of praise to our scientists. They are the true philanthropists, and what joy they must experience in their life of service to humanity.

What pleasure must thrill the heart of a man like Dr. A.D. Lloyd, as his journey through life continually brings him in contact with people whom his scientific knowledge has lifted from poverty to riches. To a man of his ripe humanity, it must be the highest degree of earthly joy to witness the supreme happiness of these people who have grown rich through their confidence in his mastery of scientific principles, and are now enjoying all the blessings which wealth can bestow.

And who are these people who have been so bountifully blessed by the work of Dr. Lloyd?

The complete answer to that question would fill volumes. There are literally myriads of them. They constitute all the people who profited from the thirty-odd oil fields which he discovered. The latest crop of them comprises those who have made fortunes in the East Texas field, Dr. Lloyd's most recent discovery. We cannot begin to name all of them in a book this size, but we can mention a few by way of illustration.

J. T. Boyce was a salesman of typewriters. He invested $50 in the East Texas field before the discovery well was brought in on geological information supplied by Dr. Lloyd. He has taken down a profit of $6,000, which is just 12,000% on his investment.

H. E. Steele received $100 for every dollar that he put into the deal and has more money coming.

Miss Ann Gordon invested a small sum and now it brings her an income of $1,500 to $2,000 per month.

C. E. Gardner and W. J. Gardner, two candy salesmen, made a sweet little investment of $1,500 in the East Texas field, and they have received from it a sweet little fortune of more than $150,000.

Mr. and Mrs. O. Currin put in the modest sum of $230 and have already taken out $87,750 in cash and oil and still have an equity of $60,000 coming to them.

Miss Florence Dahman, a stenographer, invested $40. She has collected $1,200 and has more money to come.

Miss Mae Cook, a clerk, says she made more than 100 for one, and now she is going to take a vacation.

W. J. Gamble was a policeman. For many years he pounded pavements, but he will pound them no more. He made an investment in the East Texas field, and now he can retire and live in luxury the balance of his life.

Narge Beatty, an insurance agent, invested $300 in East Texas, and has been offered $25,000 for his profit.

A. R. Hogg, Jr., a filling station attendant, invested $400, and has been offered $12,000 profit.

Charles E. Ransom, a garage mechanic, expects to educate his two boys as a result of a $35 investment in the East Texas field.

J. A. Gladdish invested $75. He has already taken down $24,100 and has enough interest remaining to pay him a large monthly income.

Mrs. Jeffers invested $300. She has taken down $11,000 and has an equity left which she values at about another $11,000.

D. M. Stern, a salesman, invested $500. He has realized $4,000 and he considers the balance of his investment to be worth $75,000.

Eugene Taber is touring Europe this summer. He invested

$500 in East Texas, from which he has realized $99,000.

Miss Phillips invested $500. She sold one-fifth of her holdings for $70,000. On that basis her entire investment is worth $350,000—just 700 for one.

Mr. and Mrs. C. A. Bodine invested $1,000. They have already taken out $35,000 in cash and oil, and Mr. Bodine estimates his remaining equity in East Texas oil investments to be worth $1,128,000.

And all this in this year of depression, when men were crying aloud that the oil business was ruined and that everything else was going to the demnition bow-wows.

The people named here are not phantoms or ghosts. They are real flesh and blood people, very much alive just now and playing a very active part on the stage of life. They all have one characteristic in common. They do not falter and hesitate when opportunity comes along. They are not doubters, but meet life with joy and hope and proper confidence.

Many of these people are from humbler walks of life, and Dr. Lloyd rejoices with them in their new-found fortune and trusts that it may bring them lasting happiness. To him they are the answer to the question, "Do oil investments pay?" To him, as to every other oil man who is in close touch with the industry, and to whom these miracles have become commonplace, these new-made fortunes are but the proof of his abiding faith that "the oil business is the greatest money-making business in the world." We do not wonder that Dr. Lloyd, after following the business for thirty years, is still fired with all the enthusiasm of a crusader and stands ready to demonstrate that more fortunes can be made.

Yes, oil investments pay, and when they do pay, it is not in any niggardly, half-hearted way. When they pay, it is with the sort of generosity that removes mortgages, pays up the life insurance, provides groceries and shelter for the rest of life, and then sends the children through college. Yes, oil investments pay.

THE LLOYD OIL CORPORATION

His knowledge of facts and conditions, such as are set forth in the preceding chapter, led Dr. A. D. Lloyd to take a very decisive step in the year 1928. In that year he organized the Lloyd Oil Corporation.

Here was a man who had spent a quarter of a century in the oil fields. He had been a pioneer in both thought and action during all those years. He had, as a geologist, pointed the way to the discovery of more oil than any living man. One could truly say that he had added more actual wealth to the world than any other man, even the world's wealthiest man. He had, of course, taken his own percentage from all these fields, but he saw the vast bulk of the wealth going to companies and individuals who had contributed nothing to its discovery. In the course of his professional work he had made money for many people and had acquired a great host of friends. He now conceived the idea of banding these friends together in the organization of an oil company which should stand as a monument to his own scientific accomplishments and as an instrument for amassing wealth for them.

"An institution is but the lengthening shadow of a man," quoth the philosopher. This truth applies no less to the institutions of business than to those of learning. Every submarine must have its periscope; the big gun requires a range finder; and the airplane's ability to combat fog and overcome navigation difficulties has been multiplied many times by the perfection of its uncanny steering instruments. So must every business have an "Eye of Knowledge" to steer its course in the way of wisdom.

In the oil business this function devolves with special emphasis upon the geologist. It falls to his lot to chart the course and make the momentous decisions which must determine success or failure for the organization. Such being the case, with a record for achievement such as his, why not "put it to the touch, to win or lose it all?"

No doubt he reflected upon the results that would have accrued had he organized such a company ten or fifteen years ago and gathered into its fold the many fields which he has discovered. Suppose such a company had taken the cream of the Caddo field in Louisiana, the Breckenridge field, the Cement field, the Newkirk field, the East El Dorado fields, Sipe Springs, South Bend, Cross-Cut and Fry fields? Suppose that to this had been added the great production of the Texas Panhandle field and the prolific salt dome fields of the Gulf coast which were discovered by Dr. Lloyd? Such wealth is almost beyond man's ability to calculate and contemplation of it staggers the imagination. Some of these fields were capable of making nearly a half million barrels per day—and they were producing when oil sold from $1 to $3 per barrel! No one can easily arrive at the sum total of such wealth, but any company that had acquired all these holdings would easily have been in the billion dollar class. And the beauty of the situation is that it would have cost very little money to acquire these properties at the time that Dr. Lloyd saw their value—in advance of the drill. That is where the big money would have been made, by buying in advance of the drill on the geological judgment of the man who has discovered more oil fields, and bigger ones, than any other man in history.

So, with Ford erecting a monument to his mechanical genius, in the form of Ford Motor Co.; with the Firestones and Goodyears perpetuating their names in commercial organizations; with Edison writing his name across the sky in forms of living light to commemorate his scientific discoveries—with these and hundreds of other worthy examples before him, why should not the man, whose scientific achievements had contributed more than any other man to the wealth and success of the petroleum industry, conceive the plan of building up a great producing oil company as a lasting monument to his pioneering accomplishments?

Such a venture would serve a double purpose. Not only would it furnish active and constructive employment for the intellect of a man who had scarcely known an idle day during his life, but it would give him the opportunity to link together, for purposes of mutual gain, the host of friends and admirers whom his driving energy and dynamic personality had attracted to him during his long career. Dr. Lloyd has made many fortunes, but he has nothing of the hoarding instinct. The making of money, not the keeping of it, is his great passion. To him it seemed unimportant that he should accumulate money, but infinitely important that those associated with him should do so.

And so the Lloyd Oil Corporation was organized—organized for these purposes of mutual interest, a mutuality which was aptly expressed by its motto—"Build Lloyd, and Lloyd will build you."

…The past year has put a fearful strain upon all business. Deflation is a cruel process and cannot be justified by any system of morality, ethics or religion known to man. Values of property have been destroyed as much so as if the torch had been applied. This has happened to farm acreage. It has happened to the corner building in our own town. It has happened still more emphatically to the oil business. Under these conditions, necessarily, much of the value has been taken from the properties of the Lloyd Oil Corporation evaluated by Rodgers, Smith & Co., on July 31, 1931. In addition to the general business depression there has been the influence of the great East Texas field. Due to its geographical location, its consequent low freight rates to the gasoline market, its low lifting cost, its high quality and its flood of production, it has, for the time being, nullified the value of practically all other oil property—this for the reason that in these respects no other oil field can compete with it.

To meet this situation, the Lloyd Oil Corporation has devoted much of its resources to the acquisition of East Texas acreage. Some of this is proven, some semi-proven, some wildcat. What its value is will be determined as it is developed and as conditions in the oil business become more stable. The officials of the company believe that it far exceeds the values which have been lost by deflation.

As this is written, the petroleum industry seems to have passed the trough of the depression. The shut-down in Texas and Oklahoma is fast clearing the atmosphere. Values are on the mend, and oil men, generally, are of the opinion that prices will soon be back to a dollar a barrel, or even better.

Should such conditions materialize, the future of the Lloyd Oil Corporation would seem rosy indeed. Not only would it derive great profits from the enhancements of its present properties, but, under the leadership of its able president, it would be in a strategic position to go forward to new victories.

In this highly complicated business of producing crude oil, intelligent direction is the winning factor. The company which does not have it cannot meet the constant siege and onslaughts nor endure the galling enfilading fire of competition. Happily for the Lloyd Oil Corporation, that organization possesses a resource which no other company can have and that resource is Dr. A. D. Lloyd himself.

Endowed by nature with a rugged physique and similarly blessed with an intellect, keen and strong and penetrating, Dr. Lloyd stands forth as the personification of man unafraid. Supremely confident in the correctness of his own judgment, he never refuses to take a stand or to declare his position on any vital question. With exalted courage he meets the issues of business and of life. His zestful spirit forever looks forward with optimism. He has no patience with those who believe that the good things in life are all in the past. For, in his innermost soul, there burns a light of faith which guides him, and a voice which whispers to him that he will discover other fields as great as Seminole and Texas Panhandle and East Texas, and that these will be for the Lloyd Oil Corporation and for the enrichment of the friends who have followed him and have been loyal to him throughout life.

THE END.

* * *

Copies of Cubage's book went out to the sucker list in late 1931 and early 1932, along with the usual promotional letters pushing stock in Lloyd Oil Corporation. Joiner, too, had been getting his name in print on the back of a Dallas self-help author and it may well be that Lloyd was following suit. In June 1932, Lloyd, perhaps also on the advice of Joiner, published one issue of what he called *Lloyd's Magazine*, featuring articles on East Texas celebrities including himself, Joiner and Malcolm Crim. It featured "full page portraits and informal photographs of Mr. Joiner and Dr. Lloyd." There are no extant copies known of that publication.

And then the roof caved in. By July, Lloyd Oil Corporation was operating under a restraining order. Investors were bringing claims and those claims were being investigated. No photos of Lloyd smiling in front of oil wells or books published to legitimize his company could eclipse what was on the company books. Lloyd's lawyer, with a flair for the dramatic trumped only by that of Lloyd himself, announced that there was a conspiracy afoot to destroy the successful company and declared further that those bringing claims were "lurking in the shadows waiting to pounce upon the carcass." Four days later, the receivership petition of a Pittsburgh widow and Lloyd investor, Mrs. Marie Blodgett, was approved. Lloyd's boundless energy and bluster were finally contained as he attended hearing after hearing. Lloyd was found guilty in lower courts and the wheels of justice kept right on grinding him down. In April of 1934, Lloyd and his business manager, Jimmy Cox, were convicted in District Court of two counts of mail fraud. The court acquitted the author of the fictitious biography of Lloyd that was used in the solicitations—Mr. Cubage.

While his appeal was pending, Lloyd kept milking the cow that had provided his succor for so long. At 65 years old in 1935, he had to figure that if he went to prison, he didn't have that long to live anyway. He was operating under another company name while the courts were preparing their case against him in 1934. And his last trick was a doozy. The Coahuila Oil & Gas Company was a Mexican corporation, so he claimed. Lloyd ran his operations out of Corpus Christi. He informed investors in the Lloyd Oil Corporation that their Lloyd stock had been converted to Coahuila stock. The Mexican interests of the Coahuila Oil & Gas Company had been recently sold to the British Anglo-Persian company for $400,000, which entitled each investor to $1,000 for each $200 they had spent with him. What glorious news for investors who had believed they'd been rooked by the illustrious doctor! The money was already in the bank, he said. But here's the punch line…

There was a Mexican tax, Lloyd said, on profits made by foreigners doing business in Mexico. The tariff had to be paid before the funds could be released from the bank. Each investor was to remit 5% of the amount he or she was to receive to Lloyd to pay the tariff. He included a Saltillo court decree dated July 25, 1934 stating as much. Oddly, this "official" Mexican document was written in very clear English. Lloyd was careful not to use the mails as a means of conveying his final scam to his sucker list. He sent his final promotion via express rail delivery and advised that remittance for the tariff could only be made by telegraph. When pressed by a consumer watchdog organization, the Mexican government and the Anglo-Persian Company denied any knowledge of or dealing with anyone by the name of A. D. Lloyd or Coahuila Oil & Gas.

When March of 1935 rolled around, Lloyd was back in court. The Ninth Circuit Court of Appeals at New Orleans was hearing his case. The facts presented indicated that Lloyd Oil Corporation did not have enough money to pay its employees at any time. Its assets were only worth $30,000, but Lloyd and Cox had valued them at $750,000. And the most damning bit of all—Lloyd and Cox had received half a million dollars from the sale of stock. Manager Jimmy Cox was sentenced to two 5-year sentences in Leavenworth with one sentence suspended. The freewheeling A. D. "Doc" Lloyd was sentenced to 3 years in the spiffy new penitentiary at El Reno, Oklahoma.

When Doc was sent up, the Lloyd family disbanded. May took their daughter Dannette out west where their son, Joe, was already living in Los Angeles trying to make it as a musician. Joe committed suicide in 1939. May and Dannette both began careers in nursing.

Son Douglas was keeping the Lloyd dream going while his father was in prison. In August of 1936, Doug attached himself as a "geological engineer" to the Mauldin Oil Company of Dallas. The gimmick? Mr. Mauldin had drilled all over the world. Mrs. Mauldin was billed as the "only woman oil well driller in the world!" For all the admiration Doug held for his father, he lacked the big man's charisma and natural ability to con people. By 1937, Doug had left the oil patch to resume a career as an insurance salesman and, later, as a collector for a Wichita Falls publishing company. He married and started a family—yes, just one family.

George "Tex" Lloyd continued his career of skating, telling saucy jokes, yodeling, playing the piano and putting on a good show to sell insurance all around the country well into his 50s.

By December 1937, Adelbert Durham Lloyd was out of prison. He drifted back to Texas where he found no warm reception. At 67 with no family, no legitimate work skills and nobody to admire his scientific prowess, he was badly bruised but not quite beaten. He looked around for a compadre and found an admirer in a green oil field hand from Fort Worth named Tom Chaney. He told Tom that he owned an oil company and needed a chauffeur to take him north to tend to important business. They wended their way north and ended up in a boarding house in Indiana on the banks of Lake Michigan in 1940. Once Tom realized that Lloyd was not an oil executive and that he wouldn't be paid for driving him cross-country, he took up selling used cars. Lloyd made his way to Chicago.

In the presence of his final wife, a 41 year old woman named only as "Marie" on his death certificate, Joseph Idelbert Durham, a.k.a. Adelbert Durham Lloyd, breathed his last in a boarding house at 6044 Ingleside Avenue in Chicago on March 29, 1941. He had spent the last 8 months of his life ill with an inflamed heart and failing kidneys. What was left of his person was removed to Mount Moriah Cemetery in Union Township, Ohio. The deeds that outlived his "rugged physique" are recorded here.

DR. A. D. LLOYD
Geologist, Petroleum Engineer, Pioneer Scientist

Portrait of Lloyd featured in a 1931 promotional book, illustrating his "impressive physique."

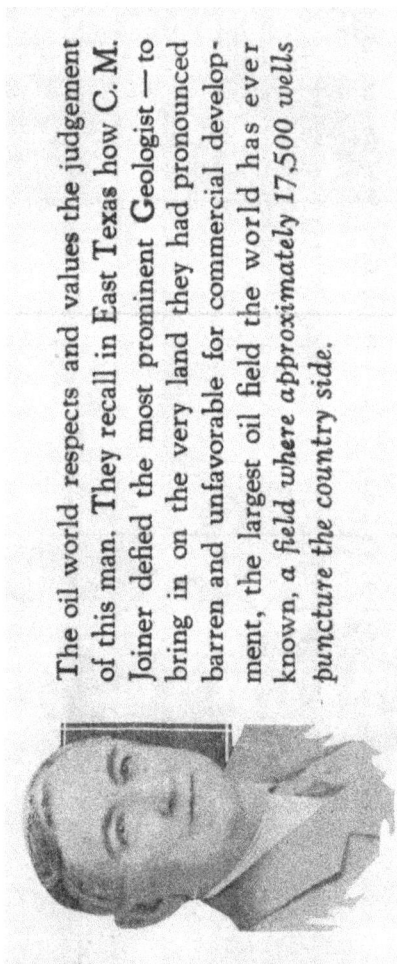

The oil world respects and values the judgement of this man. They recall in East Texas how C. M. Joiner defied the most prominent Geologist — to bring in on the very land they had pronounced barren and unfavorable for commercial development, the largest oil field the world has ever known, a field where approximately 17,500 wells puncture the country side.

From a Joiner 1936 promotional brochure, advertising lease shares in Jeff Davis County, Texas.

Lease block map which accompanied Joiner's 1932 Brewster County lease share promotion.

1931 full-page ad for the Joiner East Texas Syndicate was rife with terms and conditions on the heels of Doc Lloyd's recent trouble with the postal authorities.

DAD AFTER DAISY

Columbus Joiner's life after East Texas included more of the same shady business dealings, peppered with some high drama. He made public appearances and even acted as a guide on bus tours from Dallas to East Texas. He spoke at Writers' Club dinners and read his ode to darling Heloise. And, of course, he spent much time in court.

The 1,100 acres that he had retained in the Daisy Bradford tract had been placed in receivership. A judge allowed H. L. Hunt to withdraw 500 of those. About 200 people claimed an interest in the remaining 600 acres. But Joiner still had a nice monthly income. He blew through the money Hunt sent him each month, sometimes stopping in on him to ask for advances on the next payment. He spent money on gifts for his favorite daughter, California vacations, gifts for his beloved secretary and lots of land... for where there was land, there was the potential to sell lease shares.

Advertisements for his new venture, the Joiner East Texas Oil Syndicate, naturally featured Dad's one big achievement, accidental though it may have been. How could one go wrong investing with the man who discovered the Black Giant? Ads for the new syndicate, which ran in early 1931, were heavy on terms and conditions. Joiner didn't want to get caught in the mail fraud trap that had swallowed Lloyd. The ads stated clearly that Joiner would retain control of all monies for five years and invest them as he saw fit. Anyone sending money agreed to those terms.

In April of 1931, he changed courses. He incorporated Joiner Petroleum in Delaware on April 20 and in Texas in May. After a court appearance in August to determine whether or not the receiver could sell 102 acres of the Bradford tract (worth about $250,000) to satisfy Joiner's debts, he was asked his opinion on the martial law and proration situation in the field. Joiner could well play the role of the erudite man and Joiner could play the role of the bumpkin, depending upon which man he needed to be. 71 year old bumpkin Joiner responded, "Sure I'd rather be down there in the fields seeing what just all this means. Instead I'm kept here as a witness in this lawsuit while a bunch of lawyers wrangle over the proceeds from that oil discovery. I'd a heap rather be drilling than litigating any day…I never got a cent out of that discovery well and never expect to. I'm just here as a witness."

A few days later, Joiner let fly another one of his written works into the world. This time it was "What I Owe the Other Fellow," a short piece released to newspapers on how no self-made man is truly such; each owes his success to someone else's help. While a lofty sentiment, it was a stolen one, entirely lifted from the writing of Edgar Guest. There was not a single original thought in it. Guest's 1922 piece even lent its title to Joiner's "work." Joiner copied Guest's work almost verbatim under his own name. There were occasional minor changes but it was every bit a plagiarism if there ever was one. It ran in the Ardmore and Dallas papers and elsewhere where Joiner was interested in selling shares.

Fast forward to September 10, 1931—on that day, two seemingly conflicting news items were published half a continent away from one another. In Texas, last call went out for the Joiner syndicate receivership hearings. Up to that point, sales of property and oil had made approximately $500,000 for the benefit of leaseholders, who now numbered around 250. In Denver, the news was that sixty mining claims and two flotation mills had been sold for a figure over $1 million, half of which was cash and half of which was a stock transfer. That was big news. The big spending buyers? Dad Joiner and an associate from Chicago. 1931 finished for Columbus Joiner with notices in California and Pennsylvania papers warning the public against purchasing stock in Joiner Petroleum in those states, since the men selling it were only licensed in New York and were not permitted to do business elsewhere.

A new year brought exciting new promotions. Naturally Joiner moved away from any known or likely production. In 1932, he gravitated to Brewster County, near Alpine, and a whole mess of land in Morris County. The Morris County land was in what once was Browntown and was purchased from Clayton Browne, a Dallas real estate developer. Browne had tried to settle up a sharecropping community there but Joiner used it for another promotion. He sold it off in 1939 to a ranching outfit in exchange for little cash, but "assumption of certain obligations and transfer of other properties."

The Brewster County promotion received more of Joiner's attention. It was, after all, about as distant from East Texas, where his name and debts were known, as he could get and still be in Texas to play into Texan sentiments. He announced to the *Laredo Times* that he had blocked up 53,000 acres, and planned to spud a test well in June of 1932. He had a Houston civil engineer, J. Wharton Berlet, draw up geological reports for the promotional offerings, which included a 6-page newspaper called *Joiner Petroleum*

News, in addition to the usual sales letters and maps show-ing the future well site. The first (and only) issue of *News* included several newsy articles on proration, a full page of character and financial testimonials from friends of Joiner who also happened to be investors, an article about the sci-entific reasons why the Brewster County area was chosen, another reprinting of Joiner's plagiarized Edgar Guest piece and two pages dedicated to the promotion of Joiner Oil lease shares.

Page three of Joiner's *News* contained a coupon for those wishing to exchange other stocks of value for lease shares. For others, the cost was $5 down and $5 per month to purchase one of the 5-acre leases at the price of $5 per acre. It is very obvious that Joiner himself wrote the copy. Oc-cupying the same page was a fluff piece about what a hell of a guy the company president was. It said, in part: "Today the miracle of East Texas is known everywhere. So is the name of C. M. Joiner. Through his phenomenal string of triumphs during the past 20 years, this man has become the world-center of petroleum discussion. He is a wildcat-ter—the most sensational and most successful wildcatter the world has ever known. Mr. Joiner has probably made more poor people rich through oil investments than any man who ever lived."

But the belle of the ball in the *News* was most definitely an essay with a Joiner byline called "What Am I Striving For?" It very well encapsulated Joiner's homespun brand of truth when dealing with would-be investors and is repro-duced here in its entirety:

I am a contented man. Not seven men in a million have been as fortunate as I have been. I think I have done my meed of good in serving others, and a sense of work accom-

plished has instilled me with a peace and a calm money could not buy. I have planted trees, builded houses, drilled for oil on many a rugged hillside and in many a dale on the widespreading Texas plains. My sense of accomplishment is very great—by the work largely of my own hands I have discovered what men declare to be the greatest petroleum producing field the world has ever known. That is quite a distinction for a man who started life a farmer boy with ambitions to make his way in the world. What, then, am I striving for?

Not for money alone, to be sure. I have made money— plenty of it—sufficient for all the needs of any one man. But it isn't the money-making that thrills me; it is the zest for achievement, of things done and left behind me, of having made others happy. Nothing in life gives me greater pleasure than the feeling that I have helped others, that I have given impetus to their ambitions and comforts, to their pleasures and their happiness. I have had my difficulties and my troubles in the past, many of them, troubles and difficulties that might have staggered the courage of many a man, that might have broken the spirit and the will of men hardier perhaps than I. But these discouragements and failures did not break me, they did not harden me to the sufferings of others. I think they gave me a clearer vision of the difficulties and discouragements that others have. And I am glad that this is so. I am glad that my own difficulties did not leave me cool and calloused and indifferent to the sufferings of my fellowman. I am glad that that is so because as in memory I look back down the long winding trail I have come over to see the boyhood dreams of achievement that spread as a misty rainbow before me in those days of youth and hope, and I realize that those boyhood ambitions have not been all in vain.

I am striving, therefore, for other honors like unto those I have already received. I am striving because I have a feeling that my work is not yet complete or fulfilled. I am striving for other oil fields, and I believe, when the final chapter shall have been written into my life, that I shall be able to say, "I have not a few only, but several oil fields to my credit. From them much happiness and good has come not only to me but to others—to those who had faith in me and cooperated with me toward that end. Surely, such a life is indeed worthwhile."

In this great ambition I have spent my own money—spent it with a free hand, that in my own achievement I might make others happy, too. That ambition looms very great and very potential to me. It gives satisfaction and what more potent thing in life than satisfaction! I have known developers, men who have devoted their time to constructive effort, with whom others have associated themselves in efforts at achievement, with whom others have invested, but in my own circle of knowledge I have not known any of them who have spent their own money, their own funds, to any extent in the development in which others participated.

I say to you, and with absolute truth, that I have put my own money into this big work of developing oil fields, into this corporation, Joiner Petroleum, and into the efforts of this corporation to develop another, perhaps more than one more, big oil field. Instead of taking money out of this corporation, instead of profiting by my management of its *affairs, as many corporations have done, as you know, I have to date expended more than $100,000 in cash in the expenses of the Joiner Petroleum Corporation.*

That and more this corporation has cost me individually. Not one cent of profit have I made for myself to date from any source connected with this company. Nor do I wish to make money from it, until such time as I develop another oil field and until the profits are such that those who hold ownership in it with me, the shareholders and the owners of leases, shall also participate in rich rewards that come to those who hold ownership in a new and developing oil field.

It is fair for me to point these facts out to you—you who have purchased shares or leases with me. It is due you to know these inside details, that thereby you may better understand the purpose and aims and genuine ambitions of Joiner Petroleum. I challenge any company, any oil development project or those who head it, to show as great or greater personal expenditures in the interests of their company management and its success. Indeed the records of some managements today stand shamefully, with light of investigation and inspection, to show that corporation heads and managements have not only expended their own funds in the interests of the company and its shareholders, but have extracted money from it, often in questionable manner and often dishonestly.

When, therefore, you invest in lease property with me or with the corporation that I head—Joiner Petroleum— you may know definitely that C. M. Joiner, too, has invested in it and substantially. Thereby you know that your speculation is in good hands, that the sole objective of this company and its development projects is to prove and develop another, or perhaps more than one more, big oil field, for the benefit of all who have risked their funds with me or with my company. Speculative? Yes, but what a glorious speculation. What chances for success and what rewards with success! Along that line I shall proceed as I have proceeded, hewing to the line of honest purpose and firm determination, with loyalty to those who have trusted me and with courage and perseverance toward the one big objective – development of additional wealth from Mother Earth, the cleanest wealth, the most honest of all wealth, wealth that in the creation injures no one, takes away from not any person, but which adds to the sum of human comfort and happiness.

And if you believe that, you'll believe anything! Joiner did not wish to spend his time in far-flung Brewster County. He stayed put in Dallas and left Charles Link in charge of drilling and appeasing the natives. While the Joiner Petroleum name was being used in West Texas promotions, Dad was using the old Citizens Lease Syndicate name to sell shares and leases back in East Texas. He announced plans for wells near Overton and London. And then he sued H. L. Hunt.

As Hunt tells it, he suspected C. M. Joiner might file suit against him. The clock was ticking on the statute of limitations. Hunt had cashed in with his pipeline and the interests purchased from Joiner in 1930. So, he paid a visit to the "old wildcatter" to inquire. When the topic finally came up, Joiner teared up. He told Hunt that there was simply no way he'd ever sue because, damn it, he just loved Hunt too much. Hunt figured the reason ol' Dad was so misty was because he was lying through his teeth. He shored up his financial positions overnight. The following morning, Columbus Joiner filed suit against Hunt, Hunt's company and Hunt's financier. The suit asked for $15 million in damages and claimed that Joiner had been detained in his hotel room by Hunt on the night of the purchase causing him to miss other opportunities to sell for a higher price. The suit, filed in November 1932, was dropped the following July at Joiner's request.

Three years after the Daisy Bradford well came in, Dad Joiner was 73. His wife and most of his children still resided in Ardmore, Oklahoma, where they'd been sans Dad for about 20 years. A bombshell was soon to land squarely in the laps of Lydia and the Joiner siblings. Newspaper articles were now circulating in Texas that the grand discoverer of the Black Giant was now broke. In newspaper interviews, Joiner laughed openly at the prospect of him being penniless, claiming to be raking in $12,000 a month from the many leases he still held after selling out to Hunt. Of course, that land was all in receivership. His income was coming from H. L. Hunt's oil payments and from his own stock promotions.

In a United Press article that circulated among Texas papers on August 1, 1933, Joiner waxed nostalgic about his philanthropy following the discovery. "When we were working on the discovery well, I told one of the boys I needed a thousand dollars and I told him to hike to Overton and raise the money. I said 'You tell your friends Joiner is giving you an interest in the well, that if we get the money they'll get a similar interest.' Later I paid him $100,000 and his five or six friends who contributed the thousand dollars got a similar sum." Joiner further claimed that he'd used his money to set several drillers up as operators to help them build their fortunes.

Perhaps the most interesting part of the United Press interview was Joiner's claim that many know him as an author instead of a famous oilman. He claimed to have "two essays and a poem printed in a textbook used in colleges and universities." The truth? Not exactly as Columbus Joiner portrayed it. While Ed Laster was down on Daisy's farm poor-boying his way to the Woodbine sand, Joiner was up in Dallas scratching his writing itch. He attended Writers' Club meetings and it is probably there that he met and courted Anderson M. Baten. During the Great Depression, self-help books were as common as Hoovervilles. Baten, a Dallas writer and Shakespeare scholar, jumped aboard just as the self-help wave was cresting. Baten clearly bought in to Joiner's "I'm just a simple man who has done great things" persona. Bearing the title *The Philosophy of Life*, and published in 1930, Baten's first work was merely a collection of the writings and quotes from famous writers, thinkers and poets throughout time. That, in itself, is nothing unusual. Compilation books have always been a home library staple. What makes it odd is that Columbus Marion Joiner is in there among Napoleon and Thoreau and Confucius. The "textbook" to which Joiner referred was Baten's self-help book. The "two essays and a poem" are given verbatim here. In Joiner's florid prose, it is easy to extrapolate how the rural East Texas folks might have construed him as a man of intellect and great faith. It is only fair to note that none of these pieces appear to be plagiarized from other authors.

Wedged between Robert Ingersoll and Shakespeare, we find this Joiner gem titled "A Morning Walk":

A walk on a summer morning; the glistening of the dew;
the songs of the birds; the racing of the hare across the field;
the lowing of the herd; the soaring of the lark; the cooing of
the dove; the humming of the bee; the warbling of the bob-
o-link; the verdure of the hills; the fragrance of the flowers;
the laughter of sweet childhood; the chatter of the aged;
the distant whistle of the youth as he goes out to harvest is
a sweet blending of God's universal love. Truly, he is a fool
who says in his heart "There is no God."

A few pages later, we find Joiner musing on death, immediately above an Ingersoll piece on the same topic, though
with his typical Agnostic bent. Joiner's poem is simply
called "Death."

What is death we all so dread?
Is it the termination of a life that has fled,
Or is it a transition from care of earth,
To the realms of bliss and perpetual mirth?

If death is a transition from earth to bliss;
The meeting of loved ones long we have missed,
Why should we dread the narrow span;
The narrow divide to the spirit land?

The reason is plain when you come to think
That the laws of God are a dividing link,
And he who transgresses the laws that were given
Stands aghast at the thought of the transition.

Next to a Tennyson poem and Marcus Aurelius speech,
Joiner is back on the subject of Earth's beauty as a reflection
of God's greatness with "The Rose."

Fragrant with the sweet perfume that cannot be touched, illuminating the soul with its ethereal beauty, emblematic of God's divine love, is the rose. Who can look upon the rose, the emblem of all that is good and pure, without feeling in his heart the sacred and omnipotent power of the great Creator, who touched with His divine love the rosebud and caused it to burst forth in all its beauty, fragrant with the breath of the great Creator.

And, finally, illuminating a page otherwise occupied by Omar Khayyam and Lord Byron, is Joiner's brief treatment on beneficence called "The Gloaming."

In the gloaming the evening shadows envelop the landscape and all nature brings peace and quietude emblematic of heaven; then it is we take a retrospect of the passing day to find if that day is lost to the world, or if we have in any way rendered valid service to mankind. The poet, no doubt, had this thought in mind when he penned the immortal words: "Count that day lost whose low descending sun finds at thy hand no worthy action done."

This is certainly not what we've been lead to believe was happening in Dallas while the well was being completed. In the August 1933 United Press interview, Joiner labeled himself an "author." True enough. But the bulk of his body of work had consisted of the promotional letters he'd been writing for so very long to screw people out of their money. Unlike many promoters who used hired pens, or copywriters who specialized in his brand of selling, Joiner seems to have written all of his promos himself. One supposes he probably enjoyed it and it satisfied his longing to be successful in intellectual pursuits. L. L. James, the attorney who represented both Joiner and Hunt through the many years of litigation that followed the discovery described Joiner as "an artist at writing letters." It was his mainstay.

Immediately after he publicly scoffed at rumors that he was broke and openly talked about how much money he was making, he did what seems to be a very counter intuitive thing—he left his wife of 52 years and married his 25 year old secretary, Dea England. When asked why in one of the many hearings to follow, Joiner simply stated that he and Lydia "just aren't suit for each other." That's a conclusion he might well have reached 52 years and eight children earlier. Lydia said they had become estranged when she refused to move to Oklahoma City with him, because she knew he was living with another woman there. A month after Joiner and England married, Lydia filed suit against him, H. L. Hunt and Hunt's companies, just to cover all the bases. In her suit she claimed that she was, of course, entitled to a portion of Joiner's estate and that he was purposely moving assets around to various trustees to hide his worth from her. He had, after all, recently told her that he owned 135,000 acres of prime land in Mexico along with lots of producing properties.

And so the hearings began in 1934...and still Joiner pushed lease shares and stocks. The family fracas was picking up steam and all of the children sided with Lydia. Each would testify against C. M. Joiner, the oilman. Meanwhile, Joiner, the *auteur*, made another appearance in Anderson Baten's next self-help book, *Why Are You Standing Still?* In stark contrast to what was being said by the people who had known him their entire lives, Baten very affectionately dedicated the book to Joiner as follows:

You are aware of your responsibility to humanity. You are courage standing erect and thinking. Duty is your motive power. Your mind is in the open-air. You stand upon the principle of granite. You are an honest, loving soul. You have always begged for the opportunity to do good, to stand by a friend, to support a cause, to defend what you believed to be right.

Inside, Joiner was given the opportunity to present his principled though, humble demeanor, etc., by publishing again, under his name, "What I Owe the Other Fellow," as lifted verbatim from poet Edgar Guest in 1931. As if Baten's audacious dedication wasn't enough, the author wrote a section of the book just for his friend called "C. M. Joiner—The Moses of Texas."

Right in the back yard of Dallas, Texas, lay an undiscovered oil field, known today as the Joiner East Texas Oil Field—the largest in the world.

The major oil companies, the great independent oil companies, geologists of international fame, had condemned this territory, by core drillings, by testing with their torsion balances and their seismographs, and their other mechanical devices. The whole industry condemned this large area as being non-productive.

But, The Moses of Texas, C. M. Joiner of Dallas, cried out to the world, "OPPORTUNITY IS WHERE YOU ARE!"

Up in a small room in an office building, Mr. Joiner, without funds, without even a few pennies with which to buy food, but with an undying Vision, and with a prodigious Faith brought in the world's greatest oil field, thereby saving the homes of thousands of farmers. It lifted the mortgages off many farms and sent many boys and girls to college. It brought on a wave of prosperity that swept across the state of Texas. Yes, it went further. It influenced the oil market of the world.

And today there are thousands of wells producing millions of barrels of oil per day. The potential value runs into figures that are staggering. This field has given employment to thousands of workers.

This man, Joiner, with Vision and Faith, has proven that "OPPORTUNITY IS WHERE YOU ARE."

Back in the Dallas courtrooms, there was much he said/they said. According to Lydia Ann Joiner's testimony, Columbus came to visit her in Ardmore in the August 1933. He had a joint will for her to sign, she said. She did not have her glasses to assist her 72 year old eyes, so she asked her husband what the will contained. He said the beneficiaries would be their five daughters, four grandchildren and the loyal Miss England. Lydia thought that would be an equitable distribution of their joint property in the end. She signed. What Lydia actually put her signature to was a divorce agreement wherein she not only agreed to a divorce, but agreed to accept the house in Ardmore and $500 a month as a final divorce settlement. Joiner took that document and his secretary down to Juarez, Mexico and used it to obtain a divorce on September 7, 1933. The following day, Miss Dea England became the second Mrs. Joiner. According to Joiner's testimony, Lydia in no way relied on him to learn the nature of what she was signing. He claimed she knew it was a divorce agreement. On several occasions, he refused to state his age for the record but finally conceded that he was 74. He admitted that he had become engaged to Miss England well before asking Lydia to sign the divorce agreement.

In the midst of the Joiner divorce trial, in August of 1934, the widow Daisy Bradford, on whose farm Joiner had fallen face-first into history, died in Tyler after an appendectomy. She was 65 years old. It had been Joiner's charms that swayed her into leasing the land, her home that put a roof over the driller's head so many nights and her determination (not Joiner's) to see a well come in that made the East Texas field out of a well on her property. When the chips were down and she had to choose between Doc Lloyd and Ed Laster, she chose the real oilman. Daisy had lived out a comfortable four years following the opening of the field, in a hotel away from the action and chaos of the Black Giant that she helped to create.

Left: C. M. Joiner, age 74, during the proceedings that stemmed from his divorce to his first wife, Lydia. Right: Dea England Joiner, 26, whom Joiner married in Mexico in September 1933.

156

Back in the Joiner soap opera, daughter Fannie Joiner took the stand. As Dad's favorite daughter, she spoke the opinion of the entire family's angst regarding recent events…it was not the idea of divorce, rather it was the fact that he had married a second time without a legitimate U. S. divorce from his living wife that upset them. Lydia herself said she would have signed a divorce agreement had Dad asked, but would not have signed it if she knew her husband's intention was to marry his secretary. C. M. Joiner's pet dished on how little support her father sent home before the discovery well came in, although she frequently asked for help. In addition to helping her mother to take care of boarders, Fannie also took on a stenography job with one of Joiner's former partners to keep ends meeting in Ardmore. She made $60 a month. Dad would sometimes enclose a few $1 bills with his letters to her. After the discovery and the Hunt buyout, though, Fannie said her father sent her a fur, jewelry, a car and oil payments. But the $2,800 Ardmore grocery bill that had been mounting remained unpaid and the radio had been taken by the repo man. Fannie said she had difficulty appreciating the gifts her father sent her because, after all, he had blown through far more money than he sent home to his family.

Mattie Joiner Garrett and her husband, Roy, both took the stand against C. M., as well. Roy Garrett testified that after the discovery well came in, he was given a job by Dad Joiner. His pay was to be $1,000 per month and his duties were described as "anything he could do" for the boss. He testified that he was not paid but once and very little.

Letters written from husband to former wife were entered into evidence at this stage of the game. The man who had just been bragging in the papers about his majestic income now had to hear his own words, written privately to Lydia in late 1932, read aloud as his attorney attempted to paint a picture of a man who was on the verge of bankruptcy. "In fact I have been sorely distressed since the well came in. So many lawsuits, so many outstanding debts, and but for the fact that the newspapers were saying that I am a very rich man, I would have been ruined," he wrote to Lydia.

His attorney asked him what he meant by that last statement. "I meant that the newspapers unconsciously kept my credit good and prevented debtors from pushing me to the wall...Everybody it seems I ever knew sued me, including my two sons. I refused to fight the boys in court and settled with them on the outside." He also shocked the courtroom by stating outright that his sons had tried to "rob him." Joiner refers here to a suit filed by his son Verne in 1931 for certain oil runs that were promised him by his father. Columbus and Verne decided outside of court that the elder Joiner would pay $50,000 to the younger in lieu of the suit. Dad never paid up so, after the Dea England marriage dustup, Verne filed suit again. Joiner countersued for $10,000, claiming that his son was merely trying to cloud the title to some land in Tom Green County and was causing him damage in doing so. John Joiner, the eldest son, also sued his father when the money Dad promised wasn't forthcoming after the Hunt sale. John settled out of court for $100,000. A very angry John Joiner said during the courtroom divorce drama that he knew his father hadn't fought the settlement because a jury would have awarded him far more money.

According to the September 28, 1934 *Semi-Weekly Farm News* coverage of the trial, after discussing the unpleasantness that ensued after the discovery, "Joiner flung his right hand in front of his face helplessly and then completely broke down, necessitating a court recess." By October of 1934, the thing was mostly settled. The court ruled in Joiner's favor. Lydia would appeal it all the way to the Texas Supreme Court. Ultimately, in 1938, the final ruling was that Lydia had no claim to any portion of Joiner's estate since they were not living together when he acquired the assets and Oklahoma community property statutes applied. The original agreement (the Ardmore home and $500 a month) remained in force. Lydia did not take the issue to Oklahoma courts.

During the family upheaval, Joiner affixed his name to two small books written by Anderson Baten. One, written on the occasion of Will Rogers' death in 1935, was a 4-page pamphlet featuring a condensed biography of the celebrated author. On the cover was printed "Compliments of C.M. Joiner" and it is presumed that it was sent to his stock subscribers. Another published the following year, called *Dallas*, was likewise produced and distributed. Also four pages, *Dallas* celebrated the city with much fanfare about modernism and progress.

In January of 1936, Joiner began a promotion in Jeff Davis County. He announced to the papers that he had 100,000 acres blocked up. One test well was drilled and abandoned. But it made for a good promotion. Next to a 20 year old image of Joiner, a promotional brochure declares, "The oil world respects and values the judgment of this man" and that Joiner is "the most spectacular and successful man the petroleum industry has ever known." Naturally, it claims that the well has already been drilled to 1,200 feet and offers a generic picture of a derrick to prove it. Joiner wasn't anywhere near Jeff Davis County, though.

He was in Dallas checking the mail, cashing checks and preparing for centennial jubilees in East Texas where he was to be honored as the daddy of the grand oil field and the savior of the poor farmer. At the shindig in Henderson, Joiner was billed alongside Daisy Bradford's brother, Ed Laster and the drilling crew. Receivership of the three final East Texas leases, with interested parties numbering 150, finally ended in June.

In October of the Texas centennial year, another literary tribute was paid to Mr. Joiner by his friend Anderson M. Baten. In Baten's latest self-help tome, *The Philosophy of Success*, the section on C. M. Joiner again paints the picture of the impoverished Alabama boy whose unshakable faith and resolve made him a millionaire…the quintessential self-made man who wanted nothing more in life than to help others. One wonders how Mr. Baten escaped being an "investor" in Joiner's many schemes to view him in such a pure light. Or, perhaps more realistically, Baten was *heavily* invested and believed it to be in his own interest to promote Joiner using his writing. The high points of Baten's version of Joiner's life:

> …*Mr. Joiner says the first thing he can remember doing as a small boy was digging a hole in the earth and planting a small, black-handled knife with all the confidence in the world that it would germinate and produce many more knives. He said, "I wanted to make knives to give to the other little fellows with whom I played…"*

> *Mr. Joiner's mother passed away when he was eight years of age. She called him to her bedside as she was dying and said, "My dear boy, you have never told me a falsehood." To this day that has been deeply stamped in his mind.*

Joiner's philosophy is, "The truth is the only thing that counts…"

He has read more than 10,000 books of the world's best literature. He has in his library more than 3,000 books which represent the cream of literature. Joiner said to me, "I have opened up the greatest oil field in the world, but I would rather be the author of Gray's Elegy than to own all the oil fields in the world…" This rugged pioneer carries several books with him when he travels.

He is a self-made man. He is an accomplished scholar, yet he has spent only seven weeks in school. There is nothing that can stop this type of man…

He has probably saved more people from poverty and bankruptcy than any other living man on the American continent. He has made possible cheaper gasoline. He has proven what courage, bravery, self-reliance, nerve, mettle, grit and tenacity can do. He has always been a man of perseverance, push, industry, application and patience. His stamina, backbone and "bulldog courage" have always carried him forward. This man has a gigantic determination and an indomitable will.

While the suit brought by his first wife was crawling along, Dad Joiner took his new wife to Long Beach, California, where they spent much of 1937. Upon reading an announcement of Karle Wilson Baker's forthcoming autobiographical novel of the East Texas field, *Family Style*, he fired off a note to the Texas author and poet. He asked her to send him the very first signed copy along with an invoice—Dad Joiner never paid for anything in advance if he didn't have to. He hinted that he may be interested in many more copies provided the price of the first edition was right.

1938 brought another lawsuit Joiner's way, but this new gale was of a different caliber than those he had become used to weathering. An 84 year old East Texas gentleman by the name of George Truscott had been duped by Joiner. In 1928, Truscott had deeded 10 acres to him and Joiner promised to pay the old man $100 a month for the rest of his life. It should come as no grand surprise that Joiner paid a little here and there, and then stopped altogether. Then he offered Truscott $3,900 outright. Truscott accepted but Joiner didn't pay. So Truscott sued him in Dallas. While the attorneys were in the courtroom, Joiner lured Truscott into the jury room where the two agreed on settlement that worked well in Joiner's favor: $500 immediately, $250 more in six months, then $250 every 90 days. Truscott dropped the suit on Joiner's word. He lived two more years and likely didn't receive anything beyond the $500 Joiner gave him in 1938.

C. M. and Dea moved into a Centennial Exposition concept home in 1939. Located at 4637 Mockingbird Lane, the home was one of seven designed for and displayed during the Exposition. The Joiners purchased it fully furnished and paid $20,000 for the property. Time was running out for the old man and he slowed his pace a bit. New promotions were scarce but for one lease scheme near Del Rio being used to lure subscribers in 1940.

Dad couldn't resist the game for long. He was back to selling leases and stock in the spring of 1941, using leases in McCulloch County as bait. He was now doing business as Joiner Leasing Corporation and this would be the promotion that mattered the most to Joiner's legacy, even if he didn't realize it at the time. Joiner reached out and touched about 1,000 members of his sucker list, with his usual lease assignment offer. About fifty bought in, scattered across eighteen different states. But it only took one of the fifty to lose patience with the "most spectacular and successful man" the oil industry had ever known. Someone tipped the S.E.C. off to Joiner's schemes. An injunction was issued against Joiner Leasing for selling securities without registering with the S.E.C. The trial and appeals courts made findings of fraud on Dad's part, but refused the S.E.C.'s injunction. Joiner had, of late, been using the word "securities" liberally in his promotional materials in order to imbue them with an air of legitimacy and inspire investor confidence. In the lower courts, it was held that what Joiner was peddling (leasehold interests) was exempt from the Securities Act because they did not fall under the statutory definitions it provided.

The case, which was heard before the Supreme Court in October of 1943, brings us to Columbus Marion Joiner's most lasting contribution to history and commerce. It was the first case in which the Court was obliged to interpret whether or not oil and gas lease interests fell under the definition of "securities." The court concluded that an instrument need not be specifically enumerated in the Act to be within its purview. Joiner was in violation of the Securities Act of 1933 and the oft-cited case, *S.E.C.. v. C.M. Joiner Leasing Corp.*, established oil and gas lease shares as a form of security. The potential profits of anyone who invested rested solely on Joiner's drilling of a well or, as the Court put it in their opinion, "the undertaking to drill a well runs through the whole transaction as the thread on which everybody's beads were strung." As such, the Court interpreted Joiner's sales as falling under the "investment contract" definition of the Act.

And just like that, Joiner's long second career as a promoter came to an end. If he had to register with the S.E.C. and be subject to their scrutiny, he couldn't sell lease shares while not drilling wells on the leases. Nor could any other such con artist.

He wrote one last letter to the editor of the *Dallas Morning News* as WWII was in full swing. It may as well have been written during the Vietnam era:

> *The mind is filled and elevated by the beautiful foliage overhanging and shading the highway of East Texas. Then chug chug, go the pumps raising from the bowels of Mother Earth the stored oil where it has reposed for untold centuries. The mind cannot vision the importance of the oil produced in Texas.*
>
> *Over 200 byproducts go into the various industries of our country. To stop the oil flow would silence most of our industries including the making of our planes so vital to our transportation and the winning of peace.*
>
> *The development of oil, like farming, should be given the greatest consideration consistent with good government to the end that food and power should be abundant, both of which are most essential to the winning of the war. The wheels of commerce must be kept spinning and every acre of ground of a productive nature should be utilized.*
>
> *The race is to the swift. We must prove that we are the swiftest and most peace-loving nation in the world. If we want freedom we must grant freedom to others. To give is to receive. We can only get back what we give out. Kind words are like dew drops on the tender plant. "If we are*

true to ourselves, we cannot be false to any man." This is a time that tries our very souls. Let us meet the situation as become true Americans.

By the time he died of a heart attack just after his 87th birthday, March 27, 1947, in a Dallas hospital, his first wife Lydia had been claiming she was his widow for more than a decade. Two of his sons were dead (Bert and John), and his remaining six children had long been estranged from him. His will, scribbled on a piece of office stationary in 1940, left his entire estate to his secretary-turned-wife, Dea England Joiner. When she petitioned to probate his will in 1950, it was reported that his estate was of nominal value. What remained were a few small pieces of property worth little money. The man who had duped investors and historians alike into believing his chimerical biography of humble rags to even humbler riches slipped quietly into the canon of Texas oil lore.

HARRY B. HARTER

EAST
TEXAS
OIL
PARADE

Originally published in 1934 under the
same title by the Naylor Company

Harry Burnett Harter, born in New York State in 1899 and raised in Tulsa, served in the American Field Service in France as an ambulance driver prior to the United States' entry into WWI. After the war ended, he stayed on overseas to serve as a clerk for a military attache. His experiences in Europe and the Middle East, along with his father's oil connections back home, made Harter a natural fit for Standard Oil's overseas operations. When his career with Standard ended, he returned to the United States in the late 1920s and worked with his father in the El Dorado, Arkansas field. After the East Texas field came in, Harter moved to Tyler where he jumped into the boomtown fray and remained until his country called again. In the second World War he finished out his military career in the service of the Air Force. He died in 1970 and was interred with military honors at Fort Sam Houston National Cemetery.

Foreword

A retired East Texas banker, who has had an important part in bringing about the discovery of the world's greatest oil field, once remarked that he did not understand why someone had not written the story of oil in East Texas. One or two brief accounts of the discovery and the discoverer have appeared in national publications, but, for the most part, the writings pertaining to the East Texas oil field have been presented in the form of technical discussions in the various petroleum trade journals. Obviously, these publications fail to reach the majority of the reading public. In presenting this work, therefore, it is hoped that it will serve to supplement what has already been published with new material that will be of interest to the general reader.

The treatment of problems discussed in this work represents the author's viewpoint alone, and reflects nothing of the policies of any organization, corporation or other individual.

It is felt that interest in any story of East Texas will be enhanced by a discussion of events occurring during colonial days, and this phase of development has been touched upon briefly in order to show the derivation of land titles in the particular area in East Texas where the discovery of oil occurred.

On the eve of the celebration of the Texas Centennial, if this work can in a small measure tell of the struggles and triumphs of the citizens of East Texas in adding another colorful chapter to Texas history, then this volume will have accomplished its purpose.

Grateful acknowledgment is made to many friends who have assisted in furnishing information concerning events leading up to the discovery of oil. Their names are too numerous to mention here, but their interest in this work is largely responsible for its appearance.

Tyler, Texas
June, 1934

PROLOGUE

EAST TEXAS? Languid, lazy rows of cotton winding their serpentine way over red hills bared to the scorching sun. Motionless pines, becalmed like schooners at sea, awaiting a breeze that will carry their balsam-like fragrance afield. Cattle placidly chewing the cud in the dubious shade of bois d'arc trees. The only motion a bent figure moving slowly up the aisles of a cotton patch; the only sound a plaintive song of the negro share-cropper, low, resonant, melodious:

> *Po' farmer, po' farmer, po' farmer,*
> *Dey git all de farmer make.*
> *His clothes is full of patches,*
> *His hat is full of holes,*
> *Stoopin' down pickin' cotton*
> *F'om off de bottom boll.*
>
> *Po' farmer, po' farmer, po' farmer,*
> *Dey git all de farmer make.*
> *At de comissary,*
> *His money in deir bags,*
> *His po' li'l wife an' chillun*
> *Sit at home in rags.*
>
> *Po' farmer, po' farmer, po' farmer,*
> *Dey git all de farmer make...*

EAST TEXAS? Nothing to do but plow, plant and plow again. Darkies singing and dreaming of a heaven filled with pork chops; shuffling off on Saturday through dusty lanes to obtain a salt-pork-and-black-eyed-peas substitute from "Mistah Will, the sto' man."

Peace all over. Quiet, persistent peace; the same yesterday, the same today, the same tomorrow, the same forever.— No! This *was* East Texas!

The quiet is shattered, the peace gone. Motion invades the scene. Perspectives change. Hammer, nail and saw stroke, cringe and whine through the stillness. Wondering eyes inquire as to the meaning of a strange tower, criss-crossed with yellow pine boards, rising over the tops of the tallest trees.

A derrick; an oil derrick. The first blind, groping effort to find what lay beneath the tranquil surface of the Redlands. A faltering hand; reaching, had they known, toward the lid of Pandora's Box.

This *is* the fabulous land of East Texas! This is the place where fortunes spout from the ground hour after hour and day after day. Fourteen thousand oil derricks cast their gaunt-ribbed shadows over the hundred and twenty thousand acres. Eighty-four million feet of steel casing and tubing—sixteen thousand miles—tap the prolific oil bearing Woodbine sand where, three thousand feet below sea level, a mighty sea of oil is stored.

OIL! That slippery substance was virtually unknown a century ago, although we read somewhere that Noah caulked his ark with natural pitch, a form of petroleum, and that the streets of Valley Forge were lighted by gas emanating from fissures in the earth during the days of George Washington.

OIL! Colonel Drake, a realist from Titusville, Pa., in 1859 sank a well that produced some twenty-five barrels a day of the vile smelling stuff. This early product appeared on the market as Senaca Oil, and was for some years regarded as a panacea for numerous bodily ills of the human race.

The great Standard Oil Company of today, blessed with a Bible student somewhere in its directorate, followed a tip appearing in the Old Testament where mention is made of the occurrence of pitch on the banks of the Red Sea, and as a result of the geological explorations in this region was able to obtain concessions to valuable oil lands.

This is the promised land of East Texas! A turn of the valve on an oil well and, in an hour's time, there is enough oil in the tanks to pay for the new automobile you have been thinking about buying. The only thing about it is that the production is *prorated*. That's a new world in the oil industry. But down here in East Texas it got its first serious application. The state, for good reasons, tells you that it is illegal to produce more than a few barrels a day from your oil well. It doesn't matter what it may have cost you to drill your well—ten thousand dollars, fifteen perhaps; it doesn't matter if the well is capable of producing ten, fifteen or even twenty thousand barrels of oil daily—and many of them are—you must take your daily measure along with the owners of the other thirteen thousand nine hundred and ninety-nine wells in this mighty oil field.

This is the domain of the oil racketeer, the promoter, the gambler, the lawless. Where the money is, greed, vice and corruption spring up. With millions for stakes, the incentive has been too great for the corruptionists. They must steal the oil from the gargantuan reservoir prepared by a benevolent nature, regardless of the havoc wrought on their neighbors and on the industry. Tales of greed and graft in East Texas that put to shame the organized rackets of the metropolis gain credence as even the authorities in charge admit that tremendous volumes of "hot" (illegally produced) oil disposed of daily in East Texas while the perpetrators of the crimes go unscathed, boasting of their accomplishments.

This is just another oil field! Seas of mud when winter rains beat on the red dirt roads. Struggling mules, bogged down with tons of oil field machinery, blacksnake whips raising red welts on their miserable hides while drivers, cursing, lash them on. Mammoth trucks, laboriously crawling through the piney woods, forcing lighter traffic off the roads. The long, hot East Texas summer at last. The roads dry up and thermometers hover near the hundred-degree mark, the still air fills with clouds of choking, blinding dust. Day and night, the arteries of the oil field are filled with the restless, straining traffic. Machines must move in, more wells have to be drilled. Although the galvanic days of the new oil strike have passed, there is a constant throbbing life to be maintained.

Ah! Kilgore—a colorful oil city today. Yesterday, sprawling indolently along the right of way of the Missouri Pacific railroad, but a sleepy village. Today, awakened from its lethargy early in 1931, a typical oil country city. At the moment of its transition, Kilgore began to dress up for incoming thousands who responded to the magic cry of "Oil!" Laborers seeking employment swarmed the streets. Promoters out to exploit the simple and uninitiated filled the family-style hotels. Scouts and lease buyers for oil companies, farmers—white and black, filled the sidewalks.

Trucks, busses, cars, wagons and pedestrians alike churned through the mud-filled streets.

Overnight sprang up the inevitable shanty and tent life of the new oil town. Galvanized-iron buildings were erected along the railway sidings to house the oil field supply stores. Construction in the business district began at fever heat. Oil wells were being drilled in back yards, on town lots—anywhere the space could be found. The throbbing of machinery, the ringing of steel striking steel, the blasts of steam from hissing boilers, the churning of rotary drilling rigs, the flares of gas torches burning off the millions of feet of natural gas for which there was no market—sounds and sights familiar to oil people—now startled the populace of Kilgore.

Cafes, restaurants, lunch counters, hot dog stands appeared with the suddenness of rabbits produced from a magician's hat, and were filled day and night with eager customers. Flimsy wooden hotels were erected and rooms had to be rented in shifts. Beds still warm after the departure of recent incumbents were almost immediately occupied by others.

Add to the sights and sounds of the busy new oil city the customary odors from corrals of mules and horses, hotel disinfectants and the aroma of frying onions from hot dog stands and the senses may complete the picture of a typical oil town.

Moving picture palaces had to vie with dance halls and brothels for patrons. The driving monotony of oil field labor leaves workmen eager for diversion at the end of their "shift." Gambling, drinking and carousing flourished. The city jail was inadequate; a larger brick bastille replaced it. Soon the city boasted a new fire department, an augmented police force, a daily newspaper and a new mayor. Malcolm Crim, elected to the office in 1931, had been a merchant and banker for years. The owner of large tracts of land in the oil producing area, he has become one of the wealthiest landowners in East Texas. Mayor Crim, an ardent civic backer, has been greatly responsible for the improvements brought about in his city. Under his guidance, Kilgore was made an incorporated city and voted bonds necessary for civic improvements.

The metamorphosis brought about by the discovery of oil has not alone benefitted the towns and cities of East Texas. There's ample evidence of the new prosperity wherever one may go through the oil region. New homes have replaced the tumble-down dwellings of East Texas farmers. Magnificent new schools have been provided for the education of the children of farmers and oil field workmen. The fortunate landowners now have money in the banks, the mortgages have been paid, the children are sent off to universities. They no longer worry about crop failure and taxes—oil will take care of that!

This is the fabulous land of East Texas!

LAND:
JUAN XIMINES

It is certain that petroleum played no part in the colonization of East Texas—a part of the province of Coahuila and Texas—one hundred years ago. Nevertheless, in the month of August, 1835, less than a year before the defeat of Santa Anna and the independence of Texas from Mexico, an obscure Mexican citizen of San Antonio de Bexar, Juan Ximines, was moved to travel up to the eastern part of the province and petitioned the government for a grant of land, then almost valueless, now worth millions of dollars because of this elusive mineral—petroleum.

Juan Ximines was not vastly enthusiastic over the opportunities open to Mexican citizens in the wild eastern frontier, but the authorities in San Antonio de Bexar exerted some little pressure on native sons and daughters in order to bolster up the frontier against the ever-increasing invasion of Americans from across the border.

Extolling the advantages of the Redlands of East Texas, the servants of the government admonished: "Go east to Nacogdoches, Juan. Go east. You are doing no good here. Times have changed and there is no future in Bexar. You are now a widower and you owe it to your three niños to get some of the millions of square varas of land the government is offering. Better go now, before it is all given away, before the empresarios and los Americanos have it all."

This, of course, was propaganda emanating from the federal government in Mexico City, and was part of a program to enforce the colonization and nationalization of the eastern part of Texas. Juan's friends did not tell him that many of the colonists sent up to the frontier were convicts discharged from federal prisons in order to speed the settlement of government lands by native Mexicans. With the padre's blessing, an oxcart and two sturdy oxen, his three children and all their worldly possessions, Juan set out for the promised land.

In Austin, Texas, today reposes the petition of Juan Ximines, praying for one league of land, dated Nacogdoches, August 12, 1835, together with the field notes describing the grant. Little did Juan dream then of leaving his name to posterity, or that the same land today would be valued at twenty million dollars!

In Nacogdoches, the wheels of government moved somewhat laboriously and patience was a requisite in matters pertaining to land grants. Juan was no doubt more than once led to think ruefully of the security left behind him in San Antonio de Bexar. Life there was on a more even keel, even though possibilities for advancement were limited. The beauty and fertility of the country; the salubrity of the climate, the old Spanish missions, the usual combination of church and fortress; the marketplace where the twenty-five hundred inhabitants met for barter and trade—these were now a closed chapter in the life of Juan Ximines; he had crossed the Rubicon and entered into a new world.

Back in San Antonio, it was true: the process of decay had set in. Conditions were greatly changed from the time when it was under Spanish master. Where once the population had numbered above eight thousand, and where there

were four handsome missions of massive stone construction, some of which were capable of containing six or seven
hundred person, with interiors filled with paintings and
statues and exteriors surmounted with joyous bells, now,
in the year of Juan's departure, there was evidence that the
natural advantages of the region were sadly neglected. In
the year 1834, it was described in the diary of Dr. John
Charles Beales, as quoted in William Kennedy's *Texas*, as:

> ...one of the poorest, most miserable places in the
> country...the Indians steal the horses, rob the ran-
> chos and murder some one or two of the inhabitants
> every week. From want of union and energy, they
> tamely submit to this scourge, which all admit is in-
> flicted by a few Tehuacanos.

This change in the affairs of Bexar came about as the result of Mexico's struggle for independence from Spain in
1821, as a consequence of which there were no means of
sustaining any considerable military forces in the missions
hundreds of leagues away from the seat of the home government.

Prior to its independence from Spain, colonization
in New Spain was based on the Laws of the Indies, and
these were issued through the royal decrees of Ferdinand
V, Charles V and Phillip II, from the years 1513 to 1596.
Thus, for two and a quarter centuries there was but little
change in the plan relating to the distribution of lands.

In general, the lands open to colonization were designated by the governor of the district affected, and the
laws provided that a distinction be made between *escu-
deros* (gentlemen), *peones* (laborers) and those of inferior
rank. The size of the grants varied according to the rank

179

and merits of the grantees. We are inclined to believe that Juan Ximines was an escudero, even though he had never learned to sign his name. The hardships encountered in wresting a living from the soil in Bexar, the depredations of the Indians and the failure of the government to maintain the missions and provide educational facilities may be blamed for Juan's shortcoming. In the decade following her independence from Spain in 1821, the Mexican government had devoted more time to the problems of stabilizing the revolution-torn country than to those of colonization. True, various laws pertaining to colonization had been passed from time to time, but the response from her own citizens was negligible.

LAND:
STEPHEN F. AUSTIN &
COLONIZATION LAWS

In 1820, Moses Austin, after an unsuccessful mining venture in Missouri, attracted by the possibilities of obtaining a fresh start in a new land, applied to the Mexican government for permission to bring three hundred families to settle in Texas. Shortly after obtaining the contract from the government, and while en route to Texas to select a site for his colony, Moses Austin died of pneumonia.

Texas, and Texans, may well be proud that, on the demise of Moses Austin, destiny was to place the colonization project as an inheritance into the capable hands of the son, Stephen F. Austin. Stephen, who was singularly fitted to take charge of the new undertaking had had, beside the excellent educational advantage he received in Connecticut, several years of contact with frontier life in Missouri. A short time before the death of his father, Stephen had gone to New Orleans seeking employment. He had succeeded in obtaining an assistant editorship with the *Louisiana Advertiser* and, thanks to the patronage of a lawyer, Joseph H. Hawkins, was enabled to begin the study of law.

There is little doubt that but for the work of Stephen F. Austin in the colonization of Texas, this vast country would have remained under Mexico for many years to come. Under a less stable and persevering leader than Austin, the first colony must surely have come to grief during the first years of its struggles for existence. Austin, however, by his constant encouragement of the colonists and the excellent understanding of their problems, as well as by his tactful but persistent efforts to secure from the federal government valid titles to his concession, earned the respect of both the authorities of government and the colonists.

At San Antonio, Martinez, governor of the province of Texas, had recognized Stephen F. Austin as successor to his father and had given his approval to a schedule Austin had presented him for distributing land to his colonists. At Monterrey, three hundred miles to the south, the provincial deputation having jurisdiction over the province of Texas, replying to Martinez's report on Austin, advised that the colony might be settled provisionally, but that Austin must make his plans known to the superior government and await its decisions. The new government in Mexico City was then considering a colonization policy for Texas and the Californias, and Austin proceeded at once to the capital, arriving there on April 29, 1822.

It is well to remember that in December 1821, Austin had already introduced several families into Texas. These first colonists located on the river Brazos near the crossing of La Bahia road. When Austin found that Governor Martinez was not authorized to specify the quantity of land the new settlers should receive, he realized that this point must be settled by an act of Congress.

In Mexico City, Austin was to accomplish the almost insuperable feat of having his claims recognized and, in the midst of the other turbulent affairs of government, to secure the passage of laws that were to safeguard the future of his colony. It required almost a year of effort on Austin's part to achieve his purpose. During this time, his resources were completely absorbed and Austin had to borrow from friends in order to remain at the capital. While in Mexico City, Austin acquired a thorough knowledge of the Spanish language and personally interviewed every member of the congress. His persistence in the cause of his colony, plus his tact in handling the opponents to the government's open door policy, undoubtedly was responsible for his success. On the 14th of April, 1823, the supreme government finally confirmed the concession to Austin.

On his return to the colony, to be known as San Felipe de Austin, Austin obtained from the provisional government of Nuevo Leon, Coahuila and Texas, then in session at Monterrey, full confirmation of his powers to administer the colony, and was invested with the authority to protect the settlement from molestation by the Indians. In July 1823, the governor of Texas appointed Baron de Bastrop commissioner to superintend the surveying of the lands for the settlers in the new colony, and to issue titles, in union with Austin, in the name of the government. Austin's first colony was now an accomplished act!

Prior to the confirmation of Austin's contract, the national junta of the Mexican Empire, by decree on January 4, 1823, passed a new colonization law. Article one of this decree declared that the government of the Mexican nation would protect the liberty, property and civil rights of all foreigners who would profess the Roman Catholic apostolic religion, the established religion of the empire. As a consequence, the members of Austin's colony were obliged to adopt the Catholic religion, although it may be said that the following of its precepts did not greatly interfere with the problems of colonization.

Contractors with the government, known as "empresarios," were granted a bonus of land for each two hundred families they might introduce. To colonists whose occupation was farming the law would grant one square "labor" of land, equivalent to 177-1/8 acres. To those whose occupation was stock raising, one "sitio" or 4,428-1/4 acres was allowed. The empresarios who introduced the specified number of families were to receive three "haciendas" and two labors for each two hundred families—a mere 22,494-1/4 acres! A limit on the premium lands to be granted to empresarios was fixed at nine haciendas and six labors, and the lands not under cultivation or unpopulated within twelve years from the date of the concession were to be forfeited to the government. Was it any wonder that Texas was looked upon by the colonists, borne in on the tide of the westward movement in the United States, as a land of promise?

An ominous note in the law, apparent to prospective colonists from the United States, was the stipulation regarding slavery. After the publication of the law, there was to be "no sale or purchase of slaves which may be introduced into the Empire. The children of slaves born in the Empire…to be free at fourteen years of age." The application of this section of the colonization law doubtless acted as a deterrent to many prospective colonists from the United States who did not care to undertake the perils of homesteading in virgin territory without their slave labor. A federal act, executed April 6, 1830, designed to halt the influx of colonists from the United States, forbade the further introduction of slaves, but recognized existing slavery.

Mexico was awakening to the fact that Texas was being colonized almost exclusively by emigrants from the United

States; her own citizens were not taking advantage of the possibilities of settling in the new territory. On the other hand, fear of annexation of Texas by the United States was growing in the minds of some of the Mexican statesmen and, accordingly, a law was passed on March 24, 1825, affecting colonization in Coahuila and Texas and designed to limit the settling of frontier lands to native Mexicans only. The law specified that no additional settlements might be made within twenty frontier leagues bordering on the United States boundary line, and ten littoral leagues upon the coast of the Gulf of Mexico, except such as might meet the approbation of the executive of the Union, to whom all further petitions on the subject were to be submitted.

Following his first contract with the government, Stephen F. Austin, finding in Texas a veritable land of promise, had sent glowing reports of conditions and prospects in Texas to friends and relatives in various parts of the United States. In addition to his first contract, Austin obtained new contracts during the years 1825 to 1831 for a total of fourteen hundred additional families. Other important empresario contracts, covering lands in East Texas, were obtained from the Mexican Government by Robert Leftwich, who sought to introduce eight hundred families; Haden Edwards, eight hundred; Green De Witt, four hundred; Frost Thorn, four hundred families. Colonization was going forward by leaps and bounds.

LAND:
THE REDLANDS &
NACOGDOCHES

The lands in and around the town of Nacogdoches, extending northward to the Red River, were called the "Redlands," and as they lay within the area affected by the colonization law of March 24, 1825, virtually barred their colonization by settlers from the United States. Hence it was that these lands were included in the grants to Mexican citizens who journeyed up from San Antonio de Bexar to Nacogdoches in 1835.

Juan Ximines was but one of a considerable number of the inhabitants of Bexar, as it was called at that time, who had decided to cast his lot with the other colonists in the eastern stretches of the province. During the momentous summer of 1835, the records of the land office disclose petitions from the following citizens of San Antonio: the widow Maria Josefa Pru, with three children; Juan Vargos, with his wife and one small child; Juana Camuñez de la Viña, Maria Velarde Pena, Juana Bergar de Cadena, and others. Each of these Mexican subjects, for the reasons already noted in the case of Juan Ximines, signed with a mark "X" indicating that they could neither read nor write.

We can picture them as they set out on the hazardous journey from San Antonio to Nacogdoches. The King's Highway—El Camino Real—was not the thoroughfare the name might imply. We might conjure up visions of coaches and four, gallant horsemen with rich trappings escorting the carriages of grandees and their ladies but such was not the case. Far the greater part of the way was but a laborious trail crossing the rolling hill country east of San Antonio, at some seasons fraught with dangerous fording places in the rivers Colorado, Brazos, Trinity and Angelina.

The red mud in the bottoms caused endless difficulty and hardship on approaching the region adjacent to Nacogdoches. Bands of marauding Indians skirted the highway and took their toll from the travellers. The courageous colonists, with their slow moving oxcarts, laden with all their worldly goods, driving a few head of cattle with them, did not accomplish the journey in less than three or four weeks, although it was but a distance of two hundred miles.

The Nacogdoches Indians, from whom the city of today derived its name, were one of the confederation of twelve tribes of Texas Indians occupying the territory lying between the Trinity and Sabine Rivers. The name Texas is probably derived from the Indian word "techas," which was used by the tribes of northern and eastern Texas to denote that they were friendly to one another. In the region of East Texas dwelt the Asinai along the Neches River, the Nasoni on the Angelina, north of Nacogdoches, and the Nacogdoches themselves on the present site of the city. Like the Caddos, these were agricultural tribes, friendly to the white men. Visited in 1686 by La Salle, and in 1689 by Captain Alonzo de Leon, then governor of Coahuila, they exhibited every friendliness and asked de Leon to have some missionaries sent to them.

It is apparent that they were less zealous for the teachings of the Christian faith than for the opportunities for trading which the establishment of missions among them would bring about, as, after maintaining missions among them for fifty years, the Spanish fathers could report little, if any, headway in converting them.

The missions no doubt played a very important part in the beginning of the life of white men in East Texas, and it will be of interest here to refer to a few that were conspicuous in the vicinity of Nacogdoches. The first mission to be planted in East Texas was that of San Francisco de los Texas, in 1690. The mission was abandoned in 1693. In 1716 the Mission of La Purissima Concepcion was established near the Angelina River in Nacogdoches County, but was abandoned ten years afterward. In Nacogdoches in 1716 was founded the Mission of Guadalupe and in 1717 at San Augustine that of Dolores. These missions mark the first permanent settlement to be almost uninterruptedly occupied by European races.

In 1821, when Stephen F. Austin passed through Nacogdoches on his way to Bexar he reported that the inhabitants to the number of thirty-six had turned out to greet his party. That the growth of population in the frontier town was rapid is evidenced by an official government document dated six years later, showing there were 168 families. According to Marquis James' biography of Sam Houston, "They had cleared farms and made many improvements, roads, ferryboats, mills and gins, and houses ranging from two hundred dollars to eight hundred dollars [value]. Each of the five gins could gin two bales of cotton a day, and the past year the district had marketed two hundred bales in Nacogdoches." In 1834 the census enumerated approximately 1,350 inhabitants.

Nacogdoches, the gateway for the tide of emigration pouring in from the United States, in 1835 was a thriving community. Many of the colonists, headed for the interior of Texas, decided to remain in Nacogdoches and appropriated lands in the neighborhood without complying with the requirements of the state or federal governments. Under the colonization law of the State of Coahuila and Texas, March, 1825, the lands in this region were excluded, being within twenty leagues, or approximately forty miles, from the United States border. Thus many of the settlers were no more than squatters without any vestige of legal rights to their chosen lands.

When the lands so appropriated by settlers in the town of Nacogdoches later were included in one of the governmental empresario contracts, many of the citizens were forced to move or to submit to contracts with the empresario. In 1834 land commissioners were appointed by the government to deal with these problems and to attend to the granting and surveying of additional lands opened under the colonization laws. Native Mexicans were not excluded from obtaining grants within the twenty-league zone in the frontier.

The stretch of territory lying between the Sabine and Red Rivers was virtually a no-man's-land in which neither the authority of the United States Government nor that of Mexico was felt. Consequently it became a rendezvous for the lawless from both countries, as well as a hide-out for traders in contraband goods who found it comparatively easy to slip in to Nacogdoches and vend their wares.

In Nacogdoches, while he awaited the action of the authorities on his petition for land, Juan Ximines doubtless spent some of his idle hours in the cantinas, where Spanish proprietors provided entertainment for the wayfarer. The *alcalde*, a sort of presiding judge of the region, asked him in for a social hour. Part of the time he spent in visiting the shops of the French, Spanish or American merchants, where he priced articles that would be essential to him in clearing off his land and building his new home The alcalde's daughter, too, may have occupied his attention for some of the time. Perhaps Juan was thinking how lonely it would be, clearing off the four thousand odd acres of land up north from Nacogdoches.

In the main street of the town, idle, shiftless Indians, under the spell of civilization, drifted in from their compounds nearby and called to him to buy their seeds, corn and melons. A priest from the mission paused to pass the time of day with him and to invite him to attend mass before leaving for the north. Drab, poorly-clad soldiers from the presidio asked news of Bexar.

"Is it true that General Cos is advancing to drive out los Americanos and bring many Mexicans to Texas and Coahuila?" they asked. Juan must, perforce, shrug his shoulders and say that he has no reliable information. "Quien sabe?" he answers. "All I wish is to get this miserable business over and go on to see what I have drawn in the land lottery up north of here. They tell me the surveyors are marking the corners of my league now, but it is nearly three weeks since I arrived…I have not much money. They will have to hurry."

Unfortunately for Juan Ximines, and for the other Mexican citizens who journeyed up from Bexar during the summer of 1835, hopeful of getting settled on their grants before winter set in, affairs between the colonists and the federal government of Mexico were at the breaking point. During the time the grants to Ximines, Pru, Cadena, Peña, Vargos and the others were being surveyed, sometime in August, 1835, envoys sent by the War Party organized by the Texas colonists had arrived in Nacogdoches for the purpose of stirring the people to action. It was true that General Cos, in command of the Mexican forces, was moving toward Texas with a large army, and everywhere there was feverish activity in forming a volunteer army to repel the invasion.

Nevertheless, on August 31, 1835, Juan Ximines received the following grant from the government:

The land surveyed for Citizen Juan Ximines is located on the banks of the Angelina River, which survey begins with a stake placed near a cross road barrier bar, which turns from the large town of the Shawnees to the salt pit of Nechaz; from the said stake there are, to the south sixty-four degrees west, at two varas distance a hickory tree seven inches in diameter and to the north sixty-five degrees east, at eleven varas distance, a small china oak tree three inches in diameter. From there to the north five thousand varas to the second landmark, from which there are to be, south thirty and one-half degrees east, at eight varas distance a china oak eighteen inches in diameter and to the north, sixty-one degrees east, at eight and five tenths varas, a hickory seven inches in diameter. From there to the west one hundred varas, there is crossed a river one vara wide, at one thousand four hundred varas another (the same) one vara wide, at two thousand varas a third landmark from which there is to be, south fifty degrees east, at Five varas distance a black oak thirty inches in diameter and to the south, Fourteen degrees east, at six varas distance another twenty-nine inches in diameter. From there to the south five thousand varas to the fourth landmark from which there is to the south thirty-two and one-half degrees east, at nine varas distance a china oak eighteen inches in diameter and to the south two and one-half degrees east at nine varas distance a hickory tree five inches in diameter. From there to the East one thousand seven hundred varas as it crosses the river Striker, and to five thousand varas to the starting point, completing thus the league of land; within the said land there pertains two labors of the class temporal and the three remaining pasture land, being thus confirmed that which is found on the map accompanying it as a duplicate. Nacogdoches, August thirty-first, 1835.

William Brookfield, Surveyor
Adolfo Sterne, Translator

There follows the order from Geo. W. Smith, Commissioner, directing that the land set out in the above field notes be granted to Juan Ximines

...with the arrangement that there is due to the satisfaction of the State the sum of forty-five dollars and sixty cents, by way of satisfying Article Twenty-Two of the Colonization Law of March 24, 1825, which sums the purchaser has paid and I assert to have received it from him in conferring title on him according to my instructions and I advised him that within one year he should construct landmarks fixed and permanent in each corner of the land, that he should stock and cultivate it...

This interesting document gave good title to Juan Ximines for 4,428-1/4 acres of land, situated in what is today the southern part of Rusk County, about eight miles northwest of Henderson, the county seat. One hundred years ago the government of Coahuila and Texas collected approximately one cent an acre to cover the cost of surveying the league grant and for the stamped paper on which the deed was drawn. We are to see that Juan Ximines was unable to go into tenure of it. On March 20, 1836, the conveyance records of Nacogdoches County, Republic of Texas, disclose that Juan parted with his league of land for the sum of fifty dollars, his profit in the transaction being four dollars and forty cents.

Colonel Frost Thorn, the purchaser, also acquired the grants of Juana Berger de Cadena, Maria Velarda Peña. Francisco Castro, Juan José Albarados, José A. Cherino and others, due no doubt to the uncertainty of the position of these Mexican subjects in Texas during the war for independence. Records available indicate that most of the native Mexican citizens crossed over into Louisiana without having settled on their new East Texas lands.

Titles to the land deeded to the Mexican colonists were subsequently upheld by the Republic of Texas, and it is from the original State of Coahuila and Texas grants that titles to much of the territory comprising East Texas are derived. Unfilled empresario contracts were voided by the government of the Republic, but the titles were recognized in that portion of the contracts that had been carried out.

Colonel Thorn, himself an empresario, died on December 3, 1854, leaving many thousands of acres of land in East Texas, nearly all of which he had obtained through his contract with the Mexican government, or by the purchase of land certificates of others who obtained them from the same source. As he died intestate, the inventory of appraisement of Colonel Thorn's estate, filed in Rusk County later, shows that he left 14,875-1/3 acres in this one county alone. It is interesting to note that the value placed upon this portion of the estate was but $4,875.50. The inventory of some of the lands reveals that at that time prices ranged from one to two and a half dollars per acre, although much of it must have been valued at considerably less.

Following the trend of prices on lands located in Rusk County, we find in the county records the following transactions:

In November, 1871, S. G. Smith acquired seven and a quarter acres in the Francisco Cordova League, the consideration for which was "the building and finishing of a two-horse wagon."

In the Maria Josefa Pru League, in October 1877, J. S. Hamilton sold 325 acres to John, Eli and James Crim, accepting three notes, each payable in twenty bales of low middling cotton.

In the same survey, 1877, we read of the sale of one hundred acres or a consideration of $108 in cash, thirteen head of cattle and three five-hundred pound bales of middling cotton.

It is a far cry from the day when sixty bales of cotton could purchase 325 acres of land in the Maria Josefa Pru League. The same lands today are within the borders of the greatest oil field the world has ever known. The oil rights alone are worth from two to ten thousand dollars an acre! In Texas, cotton is no longer king. Petroleum holds the scepter.

OIL:
Exploration

"There's oil all under the state of Texas, if you only know where to find it. How about blocking up some leases in the northeastern part. There might be an extension to the Caddo pool. Don't forget Burkburnett, Mexia, Powell and all those others."

The speakers could have been any one of a dozen persons, voicing an opinion on the choicest area for some wildcat exploration for oil. Texas is a wildcatter's paradise. This was in 1919. Oil then was bringing a good price and considerable exploration was in progress. Investors were eager to put their money in wildcat wells and gamble on the discovery of oil.

McFarland, Smith and Glover, Oklahoma City oil men, were among those active in the search for new fields. They began leasing operations in Rusk County, Texas, early in 1919, acquiring from the farmers and landowners about 20,000 acres. The standard form of lease contract was employed permitting them to hold leases for ten years by paying a nominal rental each year after the first year from the date of the lease. Leases were not hard to obtain, as the majority of the landowners were anxious to have an oil well go down in their neighborhood.

It is a common practice throughout the oil country for wildcatters to lease up large areas for the purpose of promotion. Almost any area will do, although if there is some geological formation that will lend color to the undertaking it is much easier to promote the sale of acreage. Prices on this type of acreage may range from one dollar to ten or twenty, depending on the distance of the acreage from the proposed well. Many drilling blocks are obtained only by the drilling of a well at some stated point within the block. In this event, the leases usually are placed in escrow. That is to say, some local bank is made a depository for the leases, which may not be delivered to the promoters until they have fulfilled their contract to drill the well. The wildcatter depends on the sale of a part of his acreage for his profit and for the expenses of drilling the well.

The 1919 venture was doubtless but a promotion, as most of the leases were taken back to Oklahoma City and sold for a profit. Wildcat acreage, in case you do not know the phraseology, is that which lies in territory never previously drilled for oil. When the chances of getting oil are very remote, then the drilling of a well in such areas marks it as a "rank" wildcat promotion. In 1919 it was the accepted theory that East Texas was such an area.

Many of the greatest oil fields in the United States owe their development to the efforts of the wildcatters, and it is the ever-present glamor of the petroleum industry, with its Sinclairs, Marlands, Cosdens and Slicks, that lure money out of the pockets of countless thousands of business and professional people who dream of sudden riches if their wildcat comes in a gusher.

Eventually, every reservoir of oil under the earth's surface will be brought to life through the persistent efforts of those individuals or companies who engage in the business of producing oil. As the known reserves of petroleum are depleted, science will bring into play new, more advanced means of exploration. At present, geology and geophysics are in constant use in the search for new oil pools. The fact that scientific methods entail considerable expense accounts for the high percentage of failures in wildcatting operations. Nevertheless, the wildcatter will continue to wildcat, and will, by the working of the law of average, bring in some of the future oil fields.

When the Rusk County leases were disposed of in Oklahoma City to a syndicate formed for the purpose of selling stock or shares in the venture, there was a residue of a few hundred acres left over, and it remained for a veteran wildcatter to take the remnant and find the world's largest and most prolific oil field.

Columbus Marion Joiner, now known as "Dad" Joiner, of Ardmore, Oklahoma, credited with the discovery of the Duncan, Oklahoma oil field, bought these few leases in Rusk County and journeyed down to see what he had acquired. The bulk of the leases lay in the southern part of the county, between the towns of Overton and Henderson. It was a likely looking country, gently rolling hills, about one-third wooded with pines, oaks and gums, the balance given over to farms and pastures. Here for nearly a century, the people have been living on the soil, raising cotton, sweet potatoes and corn, operating gins and saw mills and dreaming that there might someday be oil.

Delighted with the appearance of the land, which struck some responsive chord within him, Mr. Joiner decided to do some leasing on his own account. Purchasing two or three thousand acres more, he revived the talk of drilling a well at an early date and even went so far as to have the timbers cut for foundations for a derrick. The timbers lay on the ground for a year or two and no well was commenced. In 1925, Mr. Joiner reappeared in East Texas and acquired some additional acreage. The actual drilling of the first well was not commenced until the summer of 1927.

Even a wildcatter may sometimes turn to the geologist for advice in locating a well. The Joiner holdings of five or six thousand acres were looked over and marked with the approval of Dr. A. D. Lloyd, who shared with Joiner the belief that the area would produce oil. The site selected by Dr. Lloyd was about two miles northwest of the site that was actually drilled. Joiner put the well on the land of Mrs. Daisy M. Bradford in order to secure the 975-1/2 acre lease for his block. The Bradford land was a part of the Juan Ximines league grant—the land the unfortunate Mexican colonist dreamed about but was destined never to see.

OIL:
The Rotary Drill

With proper financing and good luck, a wildcat well to be drilled to average depth in the Mid-Continent area of the United States should be finished up within eight or ten weeks time. The inveterate wildcatter is about the world's best optimist. If he were not, history would surely have recorded fewer strikes, and conditions in the world's greatest industry might be vastly different.

In East Texas the Woodbine sand is found at a depth of about 3,700 feet, and a well to this depth may be drilled for seven to ten thousand dollars. The machinery necessary to complete a well will cost in the neighborhood of fifteen thousand dollars. The fuel for the boilers, the replacement of bits used in drilling, lubricating oils for the engines and pumps, and countless small supplies call for an outlay of money from day to day. In wildcat operations, nearly always a long distance from the oil field supply houses, the expense is as a rule greater and delays more frequent.

Drilling a well with rotary equipment is entirely different from cable tool drilling. The rotary drill is better adapted to deep drilling and to drilling where great pressure is encountered. It is faster than any other method—and more costly.

Before a well can be commenced the ground has to be cleared off for the laying of the foundation for the derrick. Timbers, usually cut from native trees, must be prepared and rig builders called in to erect the derrick. Material can seldom be hauled in to the location for less than five hundred dollars and may easily amount to several times that amount. Slush pits must be dredged out of the ground near the derrick location for the circulation of the mud obtained in drilling. "Mud hog" pumps are installed. Racks to hold the drill pipe are built level with the derrick floor. The steam drilling engine is set on its foundation, the draw works installed and connected to the rotary table. The boilers then are located some seventy-five to one hundred feet away from the rig and a pit is dug, or a small tank erected, for the water supply.

At the top of the derrick, or crown block, a sheave is located with an arrangement of pulleys through which steel cables supporting the travelling block and pipe elevators are hung. The cables are connected to a drum operated from the engine, and the drill pipe (also called the drill stem) is caught up by elevators, hoisted by the travelling block and thus placed in position over the rotary table. The bits placed on the bottom of the drill pipe may be fish tails, diamonds or rock bits, depending upon the kind of formation encountered in drilling.

After rigging up is completed, the first operation is the drilling of a "rat hole" at one side of the derrick. This appellation is nothing more than a hole dug at a slight angle off vertical in which the top joint of drill stem, with the attached hose connection through which water is fed down the drill pipe into the drilling hole, rests when the drill pipe is taken out of the hole. This arrangement eliminates the necessity of removing the hose connection each time it is necessary to come out of the hole or to add another joint of drill pipe as the hole is deepened. The hose connection is fastened to the uppermost joint of drill pipe, the latter being named the "Kelley joint." The Kelley joint is gripped in the rotary table and, as the later is rotated, the pipe in the hole rotates with it.

The hole itself is started with a large bit, with a cutting diameter of twelve to twenty four inches. This larger hole is continued until the upper subsurface freshwater sand has been drilled through, whereupon surface casing is set and cemented to prevent the seepage of water into the hole. Usually from one hundred to two hundred feet of surface pipe is required. As drilling proceeds, water is pumped down into the hole, under pressure from the powerful mud hog pumps, and emerges at the bottom, through openings in or above the bit, causing the cuttings of clay, sand, shale, gumbo, gravel or rock formations to be washed to the surface where they flow out of the top of the surface casing into troughs which run to the slush pits. There the mud settles out and the mud hogs circulate the water back into the drill pipe, a continuous operation while drilling is in progress. The mud serves an additional purpose when gas sands with great pressure are encountered; the mud may be thickened or made heavier, and the weight of the mud in the hole prevents a possible blow-out.

OIL:
JOINER BEGINS

Not far from the Bradford farmhouse, a typical Texas dwelling with spacious galleries, surmounting a hill in the center of the thousand-acre farm, Joiner made preparations for Bradford Number One.

Heavy machinery began moving in from the highway south of the farm. In the hot summer sun, men and teams labored on the location. Soon, a yellow pine derrick arose above the fringe of willow trees growing along the banks of Johnson's Creek (the same that was referred to in Juan Ximines' grant as the River Striker). From the veranda of the Bradford home, the spectacle was a pleasing one. Sometimes the dust from plodding teams and lumbering trucks obscured the vista, but nevertheless it was a source of satisfaction to know that work had actually started.

Joiner's drilling equipment filtered in slowly from the four points of the compass and eventually settled in a place on the location. Most of the equipment, if not all of it, was secondhand. The boilers which were to furnish steam for the engine and pumps were secondhand and mismated. One was an old oil field boiler of 75 horsepower, the other a cotton gin boiler rated at 90 horsepower. The two together, we are told, could not build up more than one hundred and twenty-five pounds of steam pressure. Yet with this equipment the miracle of bringing in a new oil field eventually was accomplished!

The first well, for there were three drilled, encountered many difficulties and numerous shut-downs. The work naturally progressed slowly and financial difficulties were encountered from the start. There had been so many wild-cat promotions already, it was difficult to dispose of the acreage to advantage. Joiner had the Bradford farm sub-divided into lots of varying sizes, containing from thirty-five to fifty acres each, with an eighty acre block set out for the purpose of drilling the well.

The drill pipe employed in the drilling of the first well was old and unreliable. Frequently it stuck in the hole or "twisted off" and had to be recovered by costly fishing jobs. When once it stuck so defiantly that nothing at hand would cause it to loosen, dynamiting was resorted to and an expert called down from Arkansas to attend to the job.

Of the many stories told of the vicissitudes of the Joiner wildcatting in East Texas, the following will serve as an example: The drill pipe had stuck for a considerable time and there apparently was nothing that could be done about it. The whole countryside was intensely interested in the outcome of the development and suggestions were offered from all sides as freely as when neighbors drop in to give advice as to the treatment of one's sick child. Finally, when all other expedients had failed, it was decided to try a new method.

"All you have to do," said the adviser, "is to cut a sixty-five foot tree with a three foot base. Set this on the rotary table. Then you fix a block and tackle at the bottom of the trunk, and another at the top. Attach the travelling block and this will give you forty-nine times the pulling power you have with the ordinary travelling block." (We do not vouch for the mathematics).

It may have sounded simple, but it took some time to find a tree of suitable dimensions. A gum tree was selected, felled and trimmed and the experiment, after a great deal of trouble in getting the tree to the location, was ready. Everyone got as far away as possible, except the driller who had to turn on the steam. When the pressure was applied there was a terrific strain on the wooden derrick and, with a resounding crash, the top of the gum tree broke and went hurtling through the side of the derrick, taking with it nearly all the scantlings in that side. Fortunately no one was hurt, and the experiment proved nothing except that it would be necessary to obtain a harder wood.

About six miles south of the drilling well, an oak was purchased from a negro farmer for two and a half dollars. When it had been cut and trimmed the problem of transporting it loomed large on the horizon. However, in a country where everyone knows everybody else, someone soon thought of the proprietor of a small sawmill who would probably have the right kind of equipment for moving the log. A contact was made and a bargain arrange. The log was loaded onto four wagons and started on its way to the location.

In the first day of its journey it was hauled about half the way, but the contractor decided he was losing too much on the hauling and called the driller into conference.

"Who's going to pay for putting back all the fences I tore down and for clearing the brush away?" he demanded. Every time a right or left hand turn in the road had to be made, the fences obstructed the progress of the sixty-five foot log. The costs of hauling were mounting rapidly. When no agreement could be reached, it was decided by the drilling crew that they could get the log the rest of the way.

By the following morning, obliging farmers had loaned eight or ten teams of mules and a sufficient number of wagons. Upon reaching the spot where the log had been thrown off, they found, to their great consternation, that it had been sawed into eight-foot lengths. Someone suggested that the proprietor of the saw mill had done it, anticipating that he would not be paid for the hauling job and seeing the possibility of making eight-foot railway cross ties from the log. If such were the case, however, he was destined to be disappointed, for someone promptly volunteered to saw the logs into four-foot lengths and before nightfall the job was done. Even though the sawmill's owner was a reputed "two-gun" man and quick to resent an insult, and while it is reported that he did mount his horse and ride almost down to the well in search of revenge, nothing ever came of the incident.

The drill pipe in the hole of the first well was never loosened and the location was abandoned. The derrick was skidded over and drilling of the second Bradford well began.

Despite the enthusiasm of the East Texans, elsewhere Joiner found it difficult to arouse much interest in his project. It was necessary to sell some of the leases in order to finance the work and Joiner was called away from the well much of the time. The money came in very slowly and there were times when the well was shut down for days waiting for funds to buy much needed equipment. Sometimes there was no money for paying wages, or for settling with the grocer in Overton for foodstuffs furnished in running a cook tent for the drilling crew. Mr. Walter D. Tucker, an Overton banker, who had obtained a block of leases prior to Joiner's entry in East Texas, turned over his lease holdings to Joiner, enabling Joiner to acquire many valuable leases that he could not otherwise have had. Not only this, Tucker and his associates worked indefatigably in behalf of the Joiner drilling activity, straining their resources to the limit in order to keep the well going. During the drilling of the first two wells, in order to help Joiner in keeping expenses at a minimum. Tucker himself worked on the well with the drilling crew, while his wife superintended the cooking for them. Such was the measure of their belief in the Joiner wildcat!

Incidentally, before the completion of the third and last well on the Bradford farm, it became apparent to Joiner that his lease would expire before the well could be drilled to a sufficient depth. After almost three years of effort, it would have been disconcerting to Joiner to see the 975-1/2 acre Bradford lease slip away from him. The landowner, Mrs. Daisy M. Bradford and her brothers, H.C. and K.C. Miller, anxious to see the well completed, twice granted thirty-day extensions to the original lease.

While drilling was in progress on the third and last Joiner well, orders frequently were given to operate the well only on Sundays so that visitors from Dallas and other financial centers might see the well actually in operation and thus be induced to lend their support by buying shares in the venture.

No little credit is due the public spirited citizens of Overton, Texas, about ten miles northwest from the Joiner drilling operations, for their support and encouragement. Here it was that Joiner established a temporary headquarters for his activities during the time he sought to obtain additional leases and royalties. An obliging grocer, the same who supplied the groceries for the drilling crew, allowed

Joiner to use the back of the store as an office. Here, inside a screened-off space where the grocer piled sacks of flour and corn meal to protect them from rodents, many of the early lease transactions were enacted.

When Joiner's second well was junked and finally abandoned, his hope was at low ebb, but his friends and backers urged him to continue, promising to stay with him until the well was completed. Consequently, he moved the rig again to the third and final location. It was now in its third year on location on the Bradford farm. The drilling crew kept doggedly on. Most of them were East Texans, endowed with the rugged determination to see the job through. The head driller, E. C. Laster, kept the job going in the face of every obstacle. So firmly did he believe in the possibility of getting oil, Laster secured a number of leases for himself. Today he is the owner of a number of wells of his own.

Out of the Joiner holdings of five thousand acres, about one-fourth of this amount was syndicated and offered for sale. The plan was to sell for twenty-five dollars a one-acre interest in the syndicated holdings, with a pro-rata share in the Joiner number three well, located on an eighty-acre tract. For the purpose of fixing the potential value of the well, it was appraised at seventy-five thousand dollars. So, in the syndicate comprising five hundred acres, each twenty-five dollars invested entitled the owner to twenty-five seventy-five thousandths interest in the well itself, in addition to one five-hundredth undivided interest in the syndicate. There were three syndicates organized by Joiner, but all of them gave participation in the drilling well.

The certificates in the various syndicates were sold to friends of Joiner and friends of the landowners whose leases he had taken. In the list of subscribers appear the names of policemen, postal clerks, railway employees, bankers, merchants, waitresses, farmers and even a reputed gambler or two. Many a widow's mite and the life savings of more than one believer in East Texas were responsible for Joiner's success. Indeed, it was the support of these friends that is almost entirely responsible for the success of the venture.

In an article published in August, 1932, over the signature of C. M. Joiner, entitled "I Owe A Lot To The Other Fellow," recognition is given to his friends in the following words:

"I owe a lot to the other fellow. He has done much for me. As a matter of fact, the other fellow has made me possible. I cannot recount here all the kindly favors and helps I have received through the years; but that they are a part of me and my success, I am sure."

[NOTE: Although, as Harter notes, the article appeared over Joiner's signature, the writing was taken directly from Edgar Guest's 1922 piece of the same title.]

Harter

OIL:
The Majors Miss A Bet

A few years before the Joiner venture in East Texas, one of the major oil companies leased several thousand acres and drilled a well just outside of Henderson without getting any productive oil sand. The leases were allowed to lapse and the area was condemned, at least in the geological departments of the major companies.

Other majors had gone over the present East Texas oil area making geophysical tests without finding anything they considered of importance. Some drilling had been done by the Amerada Petroleum Company, but they, unfortunately, circled the field without getting on the inside. Sidney Powers, geologist for that company, had held to the theory that somewhere in East Texas there was a great oil field, while other geologists felt that somewhere on the East and South flank of the Caddo Arch there would be a pool. This theory was strengthened by the discovery of the Van pool in 1929, but after a number of dry holes had been drilled along the counties bordering on the Louisiana state line east of the Caddo field, it remained for the wildcatter to drill the territory which, strangely enough, had been looked on askance by most geologists.

As a precautionary measure, major oil companies employ scouts whose duty it is to report the progress of drilling in wildcat territory. In the event that indications become favorable for the discovery of oil, the companies begin buying "protection" acreage. Scouts had been visiting the scene of the Joiner drilling at intervals during the three years preceding the discovery of oil. They entertained little, if any, expectation of having anything of importance to report. Only the Humble Oil & Refining Company and the Mid-Kansas Oil Company had any holdings in the vicinity of the Joiner well at the time of the discovery of oil.

The reason for this lack of protection is probably due to the fact that the scouts, a few weeks before the completion of the well, had ceased to report it as a drilling well. Mr. Laster, the driller, has reported this oversight as being the result of having shut down in order to obtain a new string of drill pipe. Negotiations for obtaining the pipe occupied several days and the report was out that Joiner had abandoned his Bradford Number Three.

When the pipe arrived and the drilling equipment had been given a final overhauling, drilling again went on. At a depth of some 3,700 feet shale was encountered, samples of which were sent to paleontologists for analysis. The reports from the paleontologists were not favorable and both Joiner and Laster had about reached the conclusion that further drilling would be useless.

However, Laster thought he had observed a slight oiliness in the drilling mud and decided to go into the hole and take an additional core. It was this decision that was to make visible the silver lining of the Joiner venture. The core bit brought up a section of oil sand showing a considerable saturation of oil. This information, naturally, was of vast importance to all those interested in the well, and preparations were made for testing the importance of the discovery.

Shutting down the well for the day Laster removed the valuable core to Henderson and telephoned the information to Joiner in Dallas. A major oil company scout, having heard that the well had been drilling again, came to the scene and looked around the deserted drilling rig for evidence of any kind that would indicate the formation in which drilling had just ceased. He chanced upon a bucket under the derrick floor in which a few cuttings from the core bit had been caught by the driller. Finding there some oil bearing sand, he believed that the cuttings had been "salted" purposely for "bait" and that they had been obtained from a well drilling in the Woodbine sand in the Van Pool. In other words, he later told Laster, it looked suspiciously like a plant in order to promote new interest in the Joiner acreage and permit the sale of some of the leases. The word was passed around, with the result that major companies showed no enthusiasm for leases in East Texas.

After the discovery of the oil sand, in September, 1930, it was decided to make a test of the formation to determine, if possible, its potential producing qualities. A test tool was secured for the purpose and preparations got underway. First the hole had to be reamed down with a diamond bit, leaving a tapered collar in the hard shale above the sand on which to lower the test tool.

The test tool was then lowered into the hole on the end of the drill pipe, with the intention of resting it on the tapered shoulder, effectively sealing off the mud in the hole above the test tool and leaving the pressure of the gas in the sand to force the mud and other fluid below the test tool up into the drill pipe as soon as the valve was opened. The valve was opened by dropping an iron weight into the inside of the drill pipe, but apparently the test tool had not properly sealed off the mud and the test was a failure. When the pipe was brought out of the hole and uncoupled it contained only a few joints of mud. In view of the fact that the presence of an oil sand was known from the samples that had been removed in the core bit, it was decided to put casing in the well and attempt to make an oil well of it.

When it became known that the casing was set and cemented and the drilling of the plug of cement was about to commence, East Texans in the neighborhood of the well made a field day of it, coming from miles around to witness the operation. Hours before the preparations were completed, cars began arriving on the Bradford farm. The news of the impending experiment had travelled over telephones to remote corners of the county. Mail carriers on the rural routes found residents awaiting at the mail boxes for confirmation of the report. Upon receiving it, these people called their families together to prepare for the trip to the well. Lunches were packed, parasols to shield the ladies from the hot sun were hunted up and the procession began.

A crowd of eight to ten thousand people flocked to the well and prepared to stay until something happened. This was the long looked-for event in East Texas. Either success or defeat was to be the answer to their vigil. Somehow the general feeling was one of optimism, but no one could tell just why. Friends had telephoned to Mr. Joiner in Dallas to hurry down and witness the test, but he was not quite convinced that there would be anything worthwhile in the well. Finally, after a third call he consented to attend the ceremony.

It was a gala day for East Texans. "Get your soda pop and hot dogs here. Right this way, folks, for an ice cold drink!" An industrious nephew of the landowner brought out hundreds of bottles of soda water and set up his stand, doing an enviable business. In no time a hundred cases had disappeared and more were hurriedly sent for. No one seemed to mind the heat, the dust or the flies. Curious and interested people poured in from every direction. Farmers abandoned their crops, their cattle and chickens and came to watch the test. Mrs. Bradford and her brothers, with scores of friends and relatives were on the scene.

"There's Dad Joiner." A dusty Ford drew up to the well and a tired, slightly stooped, grey-haired man alighted. He was not young—almost seventy—but his step was quick and sure. Waving a hand in greeting to the many who knew him, he joined the privileged group on the derrick floor. "How does it look, Ed?" he questioned the driller. "I don't know yet, Mr. Joiner," Laster answered. "We've been bailing for hours." It was a slow process. Trip after trip into the hole, bailing out the mud to give the oil a chance to flow.

A group of scouts discussed the prospects. "I'll eat your hat if that's an oil well," one of them said.

"Well, I don't know about that," retorted another. "They have a chance to get something in the Woodbine. They checked all right on the Pecan Gap and the Austin Chalk, but the whole thing is: will they have anything in the Woodbine series? The test didn't look promising; they didn't get anything but a few joints of mud."

"Got your dice?" queried another. "You can't make an oil well out of a wildcat unless you get up a crap game. Come on, fade me somebody. Let's give the old man a break."

"Hah, a natural. Come on you seven! It looks like Joiner would get a well."

The day passed without any sign of oil. After drilling out the concrete used in setting the string of casing, the bailer was run for hours. Night came and the crowds stayed on. A few people went home for their suppers, but came back again. Many of the spectators improvised beds in their automobiles, or in the floor of their farm wagons, determined to be on hand when the well broke loose.

The bailing continued during the second day and eventually nearly all the mud was removed from the hole. Still the oil failed to flow from the well. Was it after all a failure? The crowd did not believe so and remained good natured and expectant. The indomitable spirit of the East Texans refused to give up hope.

On the third day swabbing was resorted to. The swab was run to the bottom of the casing and pulled up with it on each trip some of the mud, colored by the black fluid everyone wanted so much to see. After countless trips had been made with the swab, a change occurred.

"The fluid is rising," the driller volunteered to Joiner, who was on the derrick floor. Joiner said nothing; he was growing weary from his long vigil and the strain of many months of labor on his venture showed in the taut lines of his face. Again and again the swab was sent to the bottom of the hole.

At last there was an audible gurgling sound from the casing, coming nearer to the top all the time. A hush spread over the crowd. Something was happening. The fires in the boiler were extinguished to prevent a possible fire in case the well should begin to flow. At last a spurt of oil from the casing. The oil was flowing! A tremendous shout rose from the throng who witnessed the long hoped for spectacle.

"Oil!" they cried. "Oil!" The hilarious gathering gave vent to long-stored-up emotions. Some jumped up and down with joy, tossing straw hats high into the air to demonstrate their feelings. Mr. Joiner, witnessing the first flow of oil from his well, turned pale and leaned against the derrick for support. Then the seventy-year old veteran wildcatter turned to the driller and remarked, "I always dreamed it, but I never believed it!"

The final chapter of Joiner's drilling operations is now history known to the oil fraternity throughout the whole United States, even throughout the entire world. The discovery well, as it is known in East Texas, indicated that a good sized pool of oil had been found. There was no telling in which direction the pool would extend, but it is interesting to note that almost all of the Joiner holdings lay west and northwest of the discovery well. Overnight, the feverish leasing of lands close in began and lease hounds began to appear in Henderson and Overton as if by magic.

Joiner departed from the scene of his success and returned to his hotel in Dallas. News hounds, journalists, lease buyers and promoters were hot on his trail for information. Would he sell out and take his reward? Would he begin other wells? How much was he worth? This was the moment when many of his creditors began to feel glad that they had been willing to help him. Everything would be all right now, they thought. However, the news was scarcely out when some of them brought an action to force a receivership on his Bradford holdings, and the first fruits of his success were at once tied up in litigation.

OIL:
JOINERVILLE IS BORN

Leases which might have been bought near the well for ten dollars an acre could now be obtained only by paying five hundred to fifteen hundred dollars, since it was known that oil had been found. Somehow, the landowners sensed that the discovery was an important one. Speculators and independent operators made deals for leases on the basis of the payment of one-fourth to one-half in cash and the balance out of oil, to be paid when, as and if oil were found. For this reason some of the leases brought as much as twenty-five hundred dollars an acre.

Strange to say, the larger companies failed to reach the conclusion that this discovery would lead to a prolific field. They argued that it could be only a small oil pocket. Their geological departments had already condemned it, so, for a while, choice leases were bought up by the speculators and smaller operators.

Oil field machinery from the adjacent Texas, Louisiana, Arkansas and Oklahoma fields began moving in almost at once. Everywhere in the vicinity of the Bradford No. 3, derricks were going up and rigs began to hum. The inevitable shanty life that comes in the wake of each newly discovered oil field sprang up like a growth of mushrooms.

The well lay about one mile north of the paved highway connecting Henderson and Tyler. At the point in the highway, six miles west from Henderson, where the road turned off toward the well, lots were staked off by industrious landowners and a building boom was on. Soon cafes, grocery stores, gasoline stations, cold drink stands and drug stores were opened for business, and a thriving business it was. Sawmills were receiving the first real orders they had had in several years. Within twenty-four hours from the discovery an entire town sprang up.

The languid thoroughfare where traffic yesterday was but the placid routine travel of a stale farming community underwent a radical transfiguration. The turning point in the road became a caravanserai. Every class of vehicle, from glistening Cadillacs and Lincolns down to the lowly rattle-trap flivvers, nosed their way into the new oil empire. Along the road to the well, crudely lettered signs advertised "Rooms for Rent" and were quickly altered to read "Cots 50c a night." No one needed to ask the direction of the well, for the country road had become a solid phalanx of traffic. Amid the din of hammer and saw rose a tumult of sound from motor vehicles, as well as the crack of the whip of the four-up mule teams hauling in heavy oil field machinery. In no time, the country road became impassable. Enterprising individuals cut a new roadway and erected a one-way bridge over Johnson's Creek, leading to the well, and charged a toll of fifty cents for each passenger car and considerably more for trucks.

Up on the highway, a mile from the well, the offspring of the discovery of oil, unheralded and unchristened save for the blasphemy of countless laboring men, Joinerville was born!

In Henderson, the owners of the Justright Hotel, where once a dozen guests meant a prosperous week, were sent scurrying to the furniture dealers for extra beds. It was quickly discovered in this city that a double room would contain six or eight people. Private homes soon were filled with "paying guests" and the man who could get a decent room for fifteen dollars a week was indeed fortunate. Dining room furniture became almost useless, as family after family began to eat in the kitchen and rent out the dining room for an extra bedroom. What did it matter? All the best people were besieged with demands for bedrooms. Even the mayor had to give up a part of his home. The newcomers themselves, as in any oil boom community, set the prices for their accommodations. Could banker Jones help it, if someone begged him to take sixty or seventy dollars a month for his spare bedroom?

Everywhere signs declared the opening of this or that new subdivision. Lots were sold for ten times the value of the same space a month earlier. A new hotel was planned. The schools became so crowded that pupils attended classes only half a day. Banks profited from the effect of the oil activity as the frozen paper, cotton and farm mortgages of yesterday were paid off by farmers who, one by one, sold their oil rights. Old established law firms began to feel the pressure of the boom. Titles had to be examined before the money for leases could be delivered. Abstract and title companies were overwhelmed with orders for new abstracts of titles. Long forgotten papers were dug out of their hiding places and rushed to the county clerk's office for recording as the admonition that "a wise man records his deed" was recalled.

In the wake of the boom, came the inevitable influx of humanity. New legal talent, oil field supply agents, scouts, promoters, sharpshooters, engineers, surveyors, secretaries, adventurers and gamblers poured into the new oil land.

In Henderson, the local telephone service, with its single circuit to Dallas, was in a frenzy. A thousand or more long distance calls every twenty-four hours was not unusual. Weary operators advised: "Sorry, all circuits are busy. We will call you." One was fortunate indeed if the call might be completed within six or eight hours.

Not merely hundreds, but thousands, of the unemployed oil field workers from other parts of the country poured into the new oil domain. Hitchhikers, bums riding the blinds, families afoot, weary with their burdens, trekked into Henderson and Overton daily, forced to ask charity when they found no jobs awaiting them. As usual, the news of a new oil field had been spread by wire to newspapers the world over, and this human tide was the response.

Over at the newly completed Henderson courthouse, planned to meet the needs of Rusk County for many years to come, the records department was struggling with the receipt of hundreds of instruments hourly. Stenographers and typists from Dallas, Shreveport, Houston and other nearby cities quickly found places at the courthouse copying records, or at the abstractors' offices compiling abstracts. Fabulous wages were offered. Many a girl who had been satisfied in the city with a weekly wage of fifteen dollars found it possible to make as much, or more, in a day. Notaries public pinned badges on their hats and milled through the hotel lobbies, finding ready money on all sides. Fortunately for them the profession was not crowded, and the fees for acknowledging instruments, added to the charges they made for going out into the country in the performance of their duties, netted them comfortable sums.

In hotel bedrooms, many a deal was closed over a fruit jar of corn whiskey, in the oppressive smoke-laden atmosphere, to the accompaniment of complacent snores from other guests occupying the same rooms. Unmindful of the customary paper-thin partitions between the rooms in these hostelries, carousal and business went hand in hand; the boom was on; even the bootleggers were able to get in on the new prosperity.

HARTER

OIL:
STILL MORE OIL

By December, 1930, there were several completions in the "Joiner" field, with two dry holes chalked up that defined the eastern shore line. The first completion to follow the discovery well was the Deep Rock Oil Company No. 1 Ashby well, about one mile west. This well, showing greater pressure and volume than the Joiner discovery, was all that was needed to send drilling activities to fever heat.

By leaps and bounds the western limit of the field was extended two, three and even four miles within the first six months of the drilling campaign, a fact that astounded all observers. Nothing like it had ever been known in any field.

All over Rusk County, and in Upshur and Gregg counties north of it, locations were made and wildcat wells started. In Kilgore, twelve miles north of Joiner's activity, the Crim family, owners of two thousand acres of farm land, had long been desirous of getting a test well drilled. Before the Joiner discovery they had offered their lease to a Dallas oil man for the drilling of a well, but, at that time, the proposal did not appear attractive on land at such a distance from the discovery pool.

Coincident with the Joiner excitement, the lease on the Crim farm was given to Ed Bateman, who bad long been in the business of promoting wildcats. Bateman began the drilling of a well at once, taking advantage of the Joiner well to sell syndicated interests in his block. In New York City a banker called his friend, an oil man in Wall Street. "Say. John, do you know anything about the Bateman drilling on the Lou Della Crim land, north of the Joiner pool?"

"Sure I do, Nathan. I know you're crazy if you put any money in that proposition. Why, it's twelve miles north of Joinerville. You don't have a chance in the world. Better lay off that stuff."

"O.K., John," replied the banker, "I suppose you're right. G'bye." Nathan then proceeded to invest $750.00 in the Bateman proposition, obtaining fifteen fifty-dollar units in the venture.

In December, 1930, Kilgore rejoiced in the news of the completion of the Bateman well in the Woodbine sand, rated good for twenty thousand barrels daily, the largest well up to that moment. In New York, the banker called the oil man. "Who's crazy now?" he laughed. "I just took eight thousand dollars for my fifteen shares in that Bateman stuff." And so it goes—in East Texas.

Major companies could not ignore the terrific import of these new discoveries—the oil was there, geology or no geology. The Humble Oil & Refining Company bought the Bateman holdings of two thousand acres for a price reported to be between one and a half and two million dollars. Of course, it remained uncertain that the two fields would link up, but there were optimists aplenty who decided that they would. They were right!

Ten miles north of Kilgore came the Lathrop extension, four or five miles west of the town of Longview, with a number of tremendous producers. The amazing feature of getting oil wherever the drill went down continued until the present limits of the field were defined. The largest field in the world, measured by any yardstick that may be applied to it, came into existence within three years of furious development. One hundred twenty thousand acres of producing oil lands, extending north and south for fifty-five miles and east and west from three to nine miles. More than thirteen thousand wells had been completed by mid-summer of 1934.

Where, at first, the independents were the only true believers in the field, it did not take major companies long to make up their minds to get in the parade. Today, major companies hold approximately seventy percent or about 85,000 acres of the proven leases. The Humble Oil & Refining Company alone has fourteen hundred producing wells on its holdings.

In the wake of the drill came the inevitable growth of population within the limits of the producing area. Old towns put on new fronts; new towns formed about the nucleus of oil field camps. Gladewater, Willow Springs, London, Kilgore, Arp, Longview, Overton, Wright City, Turnertown, New London, Pirtle, Sexton City—these are but a few of the wider places in the road raised from comparative obscurity to the dignity of recognition as "towns." Of these, Overton, Longview and Kilgore are cities, with mayors, paving, water works and bond issues.

What of the greed, vice and misery of the oil field settlements? They go hand in glove with the discovery of oil. Let us look at a cross section of life in a typical oil field boom town. See that sleek new Pierce Arrow parked in front of the cafe? Well that's Jones, lousy with money. A month ago he didn't have the price of a bus ticket to Longview. How'd he get it? He played a hunch the Bateman and Joiner pools

were one and the same. Brains? Hell, no, just luck. He got
a two hundred acre lease from a colored farmer; just a little
sharp-shooting. Nothing crooked, you understand. Sold it
to a major company for a hundred thousand the day after
the Bateman well was brought in.

Those gals? Oh they're from the barrel house down there
where it says "Elite Hotel." They don't come out much in
daytime. Look kinda unnatural in daylight, but give 'em
some bright lights, a little corn likker and put some quarters in the tin piano and they'll make you think you're right
up on Fourth Avenue.

Those barefoot urchins with the faded paper flowers?
Trained to their task by slothful, indigent parents, begging
for their daily bread. Those fellows with the boxes, poles
and instruments? Engineers surveying off a lease, sweating
in the heat and dust, unmindful of the flies, gnats, chig-
gers and such like. Back in their crude field office, they
convert their data into "field notes" from which they make
the drawings and blue prints of the tracts covered on the
ground. Damn those old Spanish and Mexican land grants.
Always necessary to convert the Spanish measures to feet
and tenths. Let's see, a vara is thirty three and one-third
inches, isn't it? Trees hacked a hundred years ago, struck
down by lightning—gone. Who ever heard of marking
corners with old buggy axles stuck in the ground? Back in
God's country the lands are all sectionized. Here in East
Texas, boundaries are an unholy mumble-jumble of bois
d'arc, scrub oak, pine, hickory, walnut, sweet gum, black
gum, red gum, white gum and scores of other varieties of
trees with hack marks made generations ago. The lawyers
thrive on it. Juicy law suits are always in the offing, owing
to the confusion over boundaries and markers.

Everywhere you turn, the yellow pine and steel monuments to oil, the clatter of engines and pumps, flares burning off excess gas after the oil has passed through the separator on its way to the tanks. Steam hissing on the right and left, throbbing lines underfoot as the oil speeds on its way to tanks to be measured and sent again in bigger lines to the gulf, a hundred and fifty or two hundred miles away, where tankers lashed to the wharves or the vast receptacles at various refineries are waiting to receive it.

Wooden shacks, dirty grey tents, swarming with life, crouching beside the towering, interlacing maze of derricks. A low, ceaseless hum, an undercurrent of rhythmic sound, pierced by sharper notes as steel strikes steel. Mud, oceans of mud, belly deep as an eight-up of mules struggle and groan to move a rusty, red oil field boiler to some distant location. The cracking of the black snake whips, the defiant blasphemy of the drivers prodding the poor beasts on.

The clean smell of pine woods blistering in the sun; the pungent odor of oil. Cabbage cooking in a pot outside a squatter's tent. The stench from a corral of mules and horses nearby.

The glitter of tinsel, the scintillating flash of diamonds, the rattle of dice. "Come on you seven!" and the "Shall We Gather At The River" of the street-corner evangelist. Whites and blacks, high and low, from all walks of life, they come to make up the typical oil boom town.

OIL:
SELLING DOLLARS FOR NICKELS

Had Joiner's Bradford No. 3 been located a quarter of
a mile farther east; had the landowner failed to extend
the time on his lease for an additional sixty days; had his
friends failed him in time of need, the chances are Joiner
would have had to fold his tent and, like the Arab, steal
away into the night.

The effect of the finding of the East Texas oil pool has
been felt throughout the United States; in fact, throughout the entire petroleum producing world. In East Texas,
the discovery was regarded by the people as a special act of
providence; business was humming, new hopes and new
prospects were awakened. To the other petroleum producing states of the union, the addition of this tremendous
reservoir of oil to an already-saturated market was very
disturbing.

East Texas soon began to receive from all quarters the
blame for bringing down the petroleum price structure.
It should be remembered that, prior to the discovery of
oil in Rusk County, there was already a larger amount
of oil being produced in the United States than could be
absorbed by normal demand. The discovery of oil at new
low depths in Seminole and Oklahoma City increased
stocks on hand—stocks that were accumulating more and
more rapidly, owing to the decreased demand for refined

products and to the slowing down of our industrial machinery. Venezuelan and other South American oils were being brought into the United States in every-increasing quantities, duty free, and at prices that were below the market for oil produced at home. There is little wonder, therefore, that the oil industry in the United States was in poor condition to receive the news of the discovery of another major oil field, larger than any previously discovered in any part of the world.

Upon the completion of the Joiner well there was no outlet for the oil produced. For a few weeks, the oil was sold for fuel to fire the boilers of other wells drilling in the vicinity, and the price was fixed at one dollar and ten cents a barrel.

With the completion of the first gathering pipeline system, built by the Panola Pipe Line Corporation in December, 1930, oil from the East Texas field became a competitor with that from other fields. The first oil gathered by this pipe line was shipped by tank cars loaded at a point on the short line system of the International Great Northern Railway, connecting Henderson with Overton. At Overton the cars moved over the Missouri Pacific Railway system to the Gulf.

Other gathering pipeline systems soon were under construction. Following the Panola's example, these lines also constructed loading racks on the I.G.N. right of way. Before long an unprecedented congestion of railroad traffic resulted from the arrival of hundreds of empty tank cars destined to carry the oil out of the field to refineries in Oklahoma, Arkansas, Louisiana and Gulf ports in Texas. There was no such thing as proration then in existence, and the sole purpose of the owner of an oil well seemed to be to produce the maximum amount of oil for which a market could be found.

Railroad officials from St. Louis to Houston were besieged with telephone calls, telegraphs, letters and personal visits from irate shippers whose tank cars had failed to arrive at loading racks and whose wells had to be shut in. Passenger traffic was subjected to long delays when attempting to move through the oil belt. The tank car movement was not alone to blame for the traffic jams, as boilers, engines, pumps, pipe and other oil field equipment of all descriptions were coming in by rail as well as by motor lines. Miles of new railway sidings had to be laid before traffic could be handled in normal fashion.

Competition for pipeline connections became very keen, and considerable capital was expended in laying gathering systems in the field. Major companies were preparing to end the tank car movement of oil by the completion of numerous large trunk lines connecting the field with their refineries. It was the practice, before these lines were completed, for each producer to allow two or more pipeline companies to tie into his tanks. Then he could, naturally, play one pipeline company against another in an effort to secure better prices for his oil.

The first oil shipped out of the field via the tank car route brought sixty cents per barrel. The production from the field gradually increased to the point where the oil could not readily be disposed of at this price and a price-lowering competition resulted. Prices tumbled from sixty to fifty cents, from fifty to thirty-five and then to twenty cents. When this figure was announced everyone supposed the bottom had been reached, although some veterans of both the Smackover and Spindletop days recalled a time when oil in that field could not be moved at ten cents a barrel. All precedents were upset in July, 1931, when the price reached ten cents per barrel, with larger quantities going at five cents. It was merely the law of supply and demand at work.

A Wall Street economist worked out an astonishing comparison, whereby he proved that "selling dollars for nickels" was, in this case, far from an empty figure of speech. Assuming that he could purchase a quart of Scotch for five dollars, and that his bootlegger would use the money to purchase one hundred barrels of East Texas crude, he asked himself who had made the better investment.

Under economic scrutiny, his quart bottle of Scotch yielded only intangible assets, offset by contingent liabilities. Its heat efficiency might not be measurable in B.T.Us and its energy coefficient could only be determinable in terms of loquacity. The bootlegger's five dollar investment in crude oil revealed a treasure trove of hidden assets subjected to the ordinary refining process, the hundred barrels of crude contained seventeen hundred gallons of gasoline, twelve hundred fifty gallons of fuel oil, or the equivalent in heat and power to ten tons of coal; enough kerosene to run a farm tractor for a week and sufficient gas oil to heat an ordinary residence through a winter month.

Bringing the comparison down to the realm of concrete facts, the economist could show that adding to the original five dollar investment the cost of gathering, transporting and refining the hundred barrels would yield a profit of at least ten dollars by selling the resultant by-products at the current wholesale, f.o.b. refinery, tank car quotations at the time.

Weighing what he had bought against his bootlegger's

purchase, the economist's conclusion was that Omar Khayyam overlooked the oil industry when he wrote his famous lines:

> *I wonder what the Vintner buys*
> *One Half so precious as the stuff he sells*

HARTER

POLITICS:
OVERPRODUCTION & PRORATION

In Texas the problem of conservation of oil and gas lies within the jurisdiction of the Railroad Commission, a body composed of three commissioners whose office is elective. Control of the state's oil and gas fields is effected by means of a department known as the Oil & Gas Division. This arm of the Commission is represented in each of the main producing areas of the state by an umpire and supervisors whose duties are to see that all regulatory measures are enforced.

From time to time, as the need for new regulations arises, public hearings are called before the commissioners in Austin. Ten days notice must always precede the holding of a public hearing and in the notice the subject matter for discussion is announced so that all interested parties may have an opportunity to prepare and file their briefs.

In effect, the hearing rooms of the Commission are similar to courtrooms, with the commissioners presiding as judges. The rulings of the Commission become valid and binding upon signature by two of the three commissioners. Each commissioner is elected for a term of six years, and the chairmanship of the body is reposed in one member for a period of two years, whereupon it is passed to one of the other members.

At the time of the Joiner discovery in East Texas, the Railroad Commission was composed of Pat Neff, Chairman, C. V. Terrell and Lon A. Smith. This group naturally was watching closely the rising tide of production from the new field. Notwithstanding their attention, the daily output continued to rise until, in April, 1931, East Texas alone was producing 160,000 barrels of oil.

Prior to the dumping of this oil onto an already harassed market, the State had been receiving a gross production tax amounting to two percent of the value of the seven hundred thousand barrels coming from its other fields. With an average of one dollar per barrel as the going price of oil throughout the state prior to East Texas, the State's revenue had amounted to about fourteen thousand dollars daily; oil produced from lands owned by the state university was bringing in enormous revenues. Now, however, in spite of the 160,000 barrels coming from East Texas, the gross production tax diminished to about three thousand six hundred dollars daily. This was due to the fact that the flooded market had lowered the average price throughout the state to twenty cents a barrel.

Then it was that proration was decided upon as the panacea for the state's petroleum pains. In April, 1931, the Railroad Commission, after due hearing, adopted its first proration plan. It was decided that effective May 1st a limit of 160,000 barrels daily should be produced from East Texas wells. The order provided that 15,000 barrels daily increase would be added each fifteen days for a ninety day period.

No sooner had the opening guns of proration been discharged than the Commission was faced with forty injunction suits brought against it by producers who declared that they had contracts for much larger quantities of oil than the new proration order would permit them to produce and deliver. The federal courts granted temporary injunctions to these applicants, restraining the Commission from enforcing its proration orders in each case. This was but the beginning of a ludicrous game of battledore and shuttlecock between the Commission and a minority element of East Texas oil producers who determined to resort to any practice in order to attain their own selfish ends.

Indeed, to the many smaller independent producers, the enforcement of proration, coupled with the distressing low prices that prevailed during the first six months of 1931, appeared to threaten early ruin. With approximately twelve thousand five hundred dollars invested in each well they drilled, these operators were dependent on their oil production and sales for the wherewithal to pay the supply houses for materials furnished for drilling and equipping their leases, as well as to pay the contractors who drilled their wells for them. Their only salvation lay in higher priced oil or in recourse to injunction suits against the Railroad Commission. Many of these operators, viewing the small supervisory forces of the Railroad Commission, chose the comparatively easy plan of surreptitious violation.

From the very first, the Commission was beset with numerous difficulties which rendered the task of enforcing the proration laws almost impossible. There is little wonder that the Commission failed to cope with the flagrant violation of its orders when it is remembered that new wells were being brought in at the rate of five hundred a month and that the enforcement body consisted of a mere half-dozen of umpires and a small number of field assistants.

By June first, 1931, there were 1,000 completed wells in the field, producing and marketing 500,000 barrels of oil daily—a figure greatly in excess of the allowable production set by the Commission. The pressure of this avalanche of oil rapidly undermined the price structure of the entire industry from coast to coast and from the Great Lakes to the Gulf of Mexico. By August the price of oil throughout the Mid-Continent field had fallen from $1.07 for high gravity oil, similar to that produced in East Texas, to an average of twenty-two cents a barrel.

Injunction suits against the Railroad Commission continued to pile up daily; the dockets of the federal courts became congested to such an extent that judges instructed petitioners to pool their complaints and eliminate the necessity of separate hearings. In the interval between the date of petitioning and the date for trial, the appellants might naturally produce their wells wide open under the terms of the injunction. The law seemed to favor those who wished to cloak their unfair practices with a vestige of legality. This turn of events naturally proved discouraging to others who were striving to comply with the proration orders.

During this first eight months of 1931, numerous organizations were perfected among the right-thinking independent operators, joined by many of the land and royalty owners, who sought to combat the growing menace of overproduction, violation and thievery. These operators and their friends endeavored to assist the Railroad Commission in their task of enforcing proration; but the assistance was volunteered not so much from an altruistic feeling for the Railroad Commission as from the desire to restore normal prices through rigid proration and thus save their own investments in East Texas from utter destruction.

At first, nearly all land and royalty owners were opposed to any kind of proration—but that was before they had been educated to understand that a flooded market invariably means a cheapened market. They might readily have understood the circumstances had they been dealing in cotton, corn, wheat and the commodities of the farms. Oil, however, was a name to them synonymous with sudden riches. Established oil companies with millions at stake, joined by many of the independents, carried on a vigorous program of missionary work with the landowners in order to convert them to the idea that proration and conservation were vital factors in securing greater returns from the oil in the common reservoir underlying their lands.

This, then, was the status of proration in East Texas in mid-summer, 1931. It was soon apparent that the Railroad Commission could not effectively enforce proration without having laws with "teeth" in them to control the situation. Governor Ross S. Sterling called a special session of the legislature to do something about it. After a month's consideration, the legislature amended the existing law. It required the customary ten days' notice before the Railroad Commission could hold a hearing. In the meantime, the old law had been rescinded by the new. On August 14th, 1931, the Railroad Commission issued notice of a hearing to be held on August 25th. The purpose of the hearing was to establish a limit of production from East Texas that would prevent physical waste.

At the hearing, expert testimony was offered to show that a limit of 400,000 barrels daily might be allowed without incurring physical waste. At that time there were 1,625 wells completed in the field, each permitted to produce 185 barrels of oil daily. In reality, with hundreds of wells flowing wide open, actual production greatly exceeded the allowable figures. Expert geologists, employed by the state, offered testimony to show that the withdrawal of oil at a higher rate per day would result in the destruction of the reservoir pressure and hasten the day when all the wells would have to be pumped. The Commission thereupon entered its order fixing 400,000 barrels daily as the maximum to be produced in East Texas. Before the order could be carried out, Governor Sterling, in order to avert an actual crisis, adopted more drastic measures for controlling the new oil field.

POLITICS:
Sterling & Martial Law

With the dramatic turn of events in Oklahoma, where Governor Bill Murray proclaimed martial law and declared that all producing wells would be shut in until a price of one dollar a barrel was posted for oil in that state, the idea became popular with those advocating vigorous measures in East Texas to enforce conservation and bring about sane production without economic waste. Thus the news that Governor Sterling had placed the East Texas field under martial law did not prove so startling.

Events leading up to this move show clearly that it was indeed a wise one. A minority of East Texas producers succeeded in openly defying the rules and regulations of the Railroad Commission and flowed their wells wide open twenty-four hours a day. Most of this illegally-produced oil was sold at fifty to sixty percent less than the market price posted by the major purchasers. No small amount of it was absorbed by the numerous small refineries which early sprang up in the field. Most of the so-called refineries were but topping plants, low in cost as well as low in efficiency but numerous enough to seriously disturb the gasoline market by underselling the products of the refiners of legally produced oil. Topping plants could not possibly compete with the efficient refineries of the major companies unless they used crude obtained at figures much lower than the posted price. The deduction supports the contention that nearly all their crude came from illegal production.

For various reasons, landowners were openly dissatisfied with the results they obtained from their oil interests. Most of them had never seen an oil field before. They had expected early riches from their mineral rights. Now, however, gushers were being brought in at the rate of twenty to twenty-five a day, but the oil was selling for an average price of ten cents per barrel and their royalty interests were but a fraction of the amounts they had been led to expect.

Many of the purchasing pipeline companies delayed making payments for the oil they received for weeks or months following the first production. This was primarily due to the necessity of examining land titles and curing the numerous defects that existed. Of course, the landowner believed the pipeline companies had stolen the oil. Local newspapers dramatized the entry of the big oil companies into the field, and their efforts to secure proration, as an attempt to put the little men out of business. Some accused the major operators of stealing the landowners' oil when oil was priced at ten cents. One editor urged the landowners to padlock the wells on their farms unless they wished to have all their oil stolen at this low price, placed in storage until the supply from East Texas was exhausted, whereupon all the little men would be bankrupt and the majors could make fat profits on advancing prices.

The local press, too, was educating the home people on the geology of East Texas. Warning was given that the flush production would soon be gone, that the tremendous withdrawal of oil from the great underground pool would

not last forever and that the wells would soon cease flowing. The fact that salt water would soon take the place of oil
was predicted. With their new knowledge of the situation,
the landowners began organizing to fight against the waste
and greed so evident all about them.

First a murmur of protest, then a more determined stand
was taken by the aggrieved citizens to protect their rights in
the oil which they considered was being stolen from them.
Threats of setting fire to the wells, or of dynamiting the
tanks and lines of the aggressors, became more and more
numerous. The situation reached a crux in the early part
of August, 1931, whereupon Governor Sterling promptly
declared martial law in East Texas. On the seventeenth of
the month, every well in the field was shut in by executive order. The military forces moved to "Proration Hill,"
the location of the Railroad Commission's field forces near
Kilgore, where a large military camp was established. General Jacob F. Wolters, as commander of the Texas militia,
was called away from his duties as chief counsel for The
Texas Company to take charge of the military forces.

Although Governor Sterling was severely criticized in
some quarters for the stand he had taken, and was soon
attacked in court by injunction suits brought by some who
doubted the validity of his declaration of martial law, there
is no doubt in the minds of the majority of the oil producers, as well as the land and royalty owners, but that the East
Texas field was saved from very serious rioting, disorder
and lawlessness.

The field remained shut in tight as a drum, with not a
single well allowed to produce, until September 6th, 1931.
During the shut-down, the military forces policed the entire
field to see that the Governor's orders were observed, and

during this time the military forces worked out a system of control, in cooperation with the Railroad Commission. On September 6th, the major purchasing companies posted a price of 68 cents per barrel, with an increase on November 2nd to 83 cents per barrel. The chronic violators, who had been unafraid of the Railroad Commission's enforcement body, viewed the uniformed and armed body of the military force with considerably more respect, and for the first time since its inception proration was effectively maintained.

On September 21st, in order to keep the production within the 400,000 barrel limit set by the Railroad Commission during the August hearing, the daily well allowable was set at 185 barrels. On October 13th, the allowable was reduced to 165 barrels per well, as a result of the addition of the new wells. Up to this point, the Railroad Commission, assisted by the military forces, was in charge of the administration of its proration program in East Texas.

On the evening of October 13th, 1931, E. Constantin and J. D. Wrather, the owners of five wells in the field, obtained in Federal District Court a temporary restraining order enjoining the Railroad Commission, the Attorney General, the County Attorney of one of the counties in the military district and General Jacob Wolters, commander of the military force, "...from in any manner or method interfering with these plaintiffs in the production of oil and/or operation of their wells on the property (described in the decree) to the extent of 5,000 barrels of oil daily, from each of said wells upon said tracts afore described, and from seeking or attempting to seek to impose penalties and fines upon said plaintiffs, their agents or servants by reason of such production of oil and/or operation of wells

on the premises described in the bill of complaint." As the
nursery rhyme has it, was this not a pretty dish to set before
the King?

The restraining order quoted above did not name the
governor as a party defendant, whereupon Governor Sterling issued an executive order taking upon himself full
charge of the proration of the East Texas field. He ordered
his commanding general to prevent any well from producing more than 165 barrels a day and again, on October
29th, in order to hold the production within the 400,000
barrel limit, reduced the allowable to 150 barrels. Reduc-
tions continued until, on December 11th, the output was
fixed at 100 barrels per well.

The Constantin-Wrather case was appealed by Governor
Sterling to the United States Supreme Court, but this august body failed to uphold the action he had taken, and
martial law was ended in March, 1932, leaving the Railroad
Commission to battle alone with the violators of its orders.

The attitude of East Texas to the martial law episode,
under Governor Sterling, is ably expressed in the following
excerpt of an article published in pamphlet form by the
Texas Oil and Gas Conservation Association, a body of independent producers who desired to assist in maintaining
law and order in the oil fields of the state:

*The Governor took charge of the field and its proration
in order to prevent not only the physical waste of natural
gas and oil, but the absolute destruction of the property of the people in East Texas who owned these lands,
and to prevent rioting that appeared to be imminent in
the event the rape of the East Texas Oil Field of August
[1931] was repeated.*

Evidence of the actual danger of physical violence, riot-
ing and the taking of the law into their own hands by
owners of oil leases and royalties, was submitted to the
Governor and is substantiated by the letters, telegrams,
petitions and resolutions of numerous responsible citi-
zens and organizations in close touch with the situation.

Early in December, 1931, when it was rumored that
Martial Law would be revoked and the troops removed,
protests were filed by oil operators, royalty owners, labor-
ing men, businessmen and organizations throughout the
four counties of Gregg, Upshur, Rusk and Smith.

POLITICS:
The Racketeers Take Hold

The removal of troops from East Texas was the signal for the renewal of violations throughout the field. Again the Commission was unable to effect proper control of the situation. By June, 1932, the production from East Texas had been limited to 325,000 barrels of oil daily, this figure being taken to represent the limit of production that might be allowed without physical waste. Nevertheless, amounts variously estimated to run from 100,000 to 350,000 barrels daily in excess of this figure were finding their way into the market with devastating effect.

While there were a half-dozen modern, efficiently constructed refineries in operation in East Texas, the number of topping and skimming plants had increased to nearly fifty. These plants continued to receive all the oil they required at prices far below the posted market price for oil. Naturally, at the low prices they paid for crude, these refineries could only be receiving supplies from operators who would sell them amounts greatly in excess of the allowable set by the Railroad Commission. Methods of checking these plants were difficult, as most of the oil was run to them in a secretive underhand manner, through hidden lines and from sources unknown to the Railroad Commission employees.

Stolen oil was hauled from the wells at night. Tank trucks delivered it to topping plants; the converted oil, in the form of low-grade gasoline, was, in turn, distributed by fleets of trucks which virtually commanded the principal highways leaving the oil field for cities and towns within a radius of one hundred and twenty-five miles or more. Practically none of this product had paid the state gasoline tax; the plants manufacturing it kept no visible records of their crude purchases. Any attempt to trace the source of the oil going to these plants ended in failure.

Employees of the Railroad Commission were repeatedly turned away from the small refining plants in the field with shotguns and threats of violence. At one of the plants, employees of the Commission had demanded to be allowed to gauge the oil in a 55,000 barrel tank, but were sent off the premises. The investigators promised to return armed with warrants from the county authorities that would make it incumbent on the owners to give them access to the tank. Upon returning on the following day, the investigators found themselves blocked again. The owners had cut down the steel stairway to the top of the tank and there was no way to scale the tank without considerable delay. Of course, in the time required for the authorities to find a way to take the gauges, the owners had ample time to run out the thousands of barrels of illegally produced oil that had been in storage.

Shady practices have doubtless been resorted to in almost every other oil field in this country, but none of the old tricks will compare with the ingenuity of the East Texas oil racketeer.

Up in the northern extension of the field, a racket developed and flourished for a number of months before the authorities were convinced of any irregularity. A well was drilled to the regulation depth of some 3,750 feet to the Woodbine sand, the producing sand in the East Texas field. During the drilling of a second well on the property, it strangely began to produce oil when the drilling had progressed only to some 1,200 feet. There was no mistaking the fact that a considerable body of oil had been encountered, for the well began flowing oil of similar gravity to that produced from the Woodbine horizon. The operators promptly seized upon the situation to turn it to profit. They completed the well at a depth of 1,200 feet and reported to the Railroad Commission that they had discovered a new producing oil sand. They requested the Commission to allow them to produce their wells in the shallow sand to capacity, on the grounds that they were not a part of the regular East Texas producing area. The Commission permitted them to flow their wells without the restraint of proration.

What really had happened was that, in the first well drilled, there had occurred a rupture of the casing at the 1,200-foot level, at a point in the subsurface stratum where there existed a well defined water sand which extended throughout the northern end of the field. The oil from the Woodbine sand, finding an outlet in the casing at this point, was flowing into and saturating the water sand at the upper level. When the second well encountered the water sand it had been sufficiently saturated to begin to flow into the open hole.

This ingenious method was employed in the drilling of several wells, and the idea was applied to other leases in the same area.

Naturally, to successfully produce oil from the so-called shallow sand, it was necessary to have a deep well producing from the regular Woodbine sand, and to rip open the casing where it passed through the water sand so that the production would flow through this sand into the shallow wells that were drilled near it. After several months, the Railroad Commission became aware of this subterfuge and prohibited the drilling of any more of these pseudo-shallow wells.

Bypassing was the favorite method of getting excess oil out of the flowing wells. The operator would lay a secret, buried line either from his storage tanks or from a line leading from his producing well to the line of some refinery or pipeline. The regular allowable production would be run into the tanks daily, but at the same time, by opening the valve from the flowing well through the bypass, any amount of oil could be run to the surreptitious purchasers undetected by the investigators. The Railroad Commission employed a large number of men whose sole duty was to uncover hidden lines and bypasses; scores of such devices were brought to light through the efforts of this body, but the task of digging up the soil around all the wells of suspects in the field would have called for an army of workmen which the Railroad Commission could not hope to employ with the limited funds at its disposal.

One of the most defiant gestures in the history of the East Texas field was that of a producer who resisted all efforts of the authorities to investigate his well or to shut it in. This operator built a concrete blockhouse over the controls of the well, making it almost impregnable. For months the amount of oil taken from this well was unknown.

Bypass hunters, in another part of the field, on the outskirts of the town of Kilgore, had searched persistently for a bypass line from a certain well which they were confident was engaged in running excess oil. They suspected that the valve which controlled the running of the oil through the bypass to a refinery nearby would be found in the owner's house. Whenever they called at the house for the purpose of inspecting it, a servant would inform them that her mistress was in the bath and could not see them. Eventually, by some strategy, access to the interior of the house was obtained and the elusive bypass was found to be controlled by a valve installed behind the bathtub.

On a large lease in the Joiner area strongly suspected of running enormous quantities of excess oil, no investigator had ever been able to come on the lease at any hour of the day or night and find any of the wells flowing. The regular daily allowable oil was turned into the tanks and run to a pipeline company, yet the investigators were certain that excess oil was run during the night. It developed that the owners of the lease stationed a look-out in the top of one of the derricks. Upon seeing the headlights of a car approaching, it was the duty of this employee to come down from his vantage point and shut off the flowing wells before the investigators could reach the property. The owner of an interest in the property, who suspected the operators of some irregularity, succeeded in having a receiver appointed, at which time the above mentioned facts were brought to light.

In one instance, investigators for the Railroad Commission found that one well was being made to serve in the place of three. Naturally this would mean a large saving in drilling expense to the operator who succeeded in adopting the plan. First, a well would be completed in the usual manner. Then the drilling machinery would be moved onto a new location and the drilling of the next well commenced. Instead of actually drilling, the machinery would be kept in motion, the usual noises would take place, steam was maintained in the boilers and other indications of drilling carried out. After setting the surface casing, a hundred feet or so of ten or twelve-inch pipe, drilling stopped—the rest was merely a pretense.

When sufficient time bad elapsed to have actually completed the well, the controls, valves and other finishing details were added to the well and it was reported to the Commission as a completed well. In order to get the production into the faked hole, a line connected into the first well was laid underground into the second well and additional oil run through it as though it were the regular allowable production from the second well. The ingenious operator could thus save from fifteen to twenty thousand dollars, the cost of drilling two additional wells, and at the same time have the advantage of running the allowable production for all three wells from the one producer. Some talkative employee on the make-believe wells brought about the discovery of this sham and the authorities took punitive action against the offender.

With such unusual practices prevalent in the East Texas oil field, many of the major companies employed special investigators to check the field at night in order to uncover some of the underhanded thievery that existed. This was done partly for their own protection, as it was not uncommon for the owners of bypasses and hidden lines to locate the controls on some adjacent property. Landowners became more alert and watched the operators of wells on their farms in an effort to discover any possible overproduction of oil, as in most cases the landowner never received any accounting from the operator for oil produced in excess of the allowable figures.

Reports of specific cases of violations poured into the offices of the Railroad Commission, supported by affidavits of proof. The procedure then was that the Railroad Commission turned over the complaints to the Attorney General's department with the request that the state file suits against the violators. Because of the lack of adequate laws to handle the situation, and also because of the numerous injunctions allowed in federal courts, which tied the hands of the Attorney General's department, very few of the aggressors were ever brought to trial. As a consequence, violations continued rampant.

Eventually new legislation was sought and obtained. Some improvement resulted, but at no time since the beginning of proration in East Texas has there been a complete adherence to the orders of the regulatory body except during the period when the military forces had charge.

It is not possible here to say that proration has been a success, although it is certainly evident that, for the stabilization of the industry, some sane plan of regulation is necessary. Proration itself is not at fault; as a means of conservation of a tremendously valuable natural resource, it is invaluable, when and if properly and efficiently applied to a flush field. The difficulty is, and in East Texas always has been, to find a body both capable and competent to enforce the necessary regulatory measures.

HARTER

POLITICS:
The Feds Arrive

In September, 1933, the petroleum industry began to operate under the "Code of Fair Competition for the Petroleum Industry." Naturally, no code of fair competition could tolerate excess oil, and producers in East Texas looked for an early eradication of this evil.

A large force of federal investigators opened their headquarters in Tyler and began to take steps to eliminate excess oil. The force entering upon this new work suffered from inexperience in meeting the problems confronting it, and for some months there was no apparent change in the situation. Federal employees scoured the field at night, searching for bypasses and other signs of illegitimate activity, but violations continued in practically the same tenor as before. Figures from reliable sources indicated that, in the closing months of 1933, there still were sixty to seventy thousand barrels of illegal oil produced daily in East Texas. One may well ask why this should be, when both the Federal and State authorities had taken so determined a stand to erase all evil practices from the field.

As early as July, 1933, before the adoption of the Petroleum Code, President Roosevelt issued an executive order prohibiting the interstate or foreign transportation of illegally-produced oil. Figures available through the offices of the Railroad Commission at that time estimated that there was some 160,000 barrels of oil being produced in East Texas in excess of the daily allowable. Other estimates placed the figure as high as 225,000 barrels. The President issued the order as the result of a petition presented to him by counsel for the Independent Petroleum Association of America. The petition was supported by the Railroad Commission of Texas, all three of whose commissioners attended a meeting in Washington to state their position. It was felt that this action had to be taken to prevent the breakdown of proration in East Texas prior to the adoption of the Petroleum Code, then in the course of preparation. This order may have served its purpose in preventing the shipment of illegally produced oil out of the state, but it in no way prevented the traffic in such oil within the state.

The oil fraternity therefore looked with great expectancy to the federal investigators of the Department of the Interior to put an end to the large interstate shipments of illegal oil. This end might have been accomplished but for the efforts of a small group of chronic violators in East Texas, who obtained an injunction in the Federal Court for the East Texas district from Judge Randolph Bryant. By the terms of the injunction, the federal investigators were prevented from having access to the complainants' refineries or their records. The case, known as the "Panama" case was filed on the grounds that these operators never had signed the Petroleum Code and should not be subjected to any of the provisions of the code. The plaintiffs, in their pleading, further stated that their business was entirely interstate commerce and, therefore, would not come under the

President's executive order covering interstate shipments of illegal oil.

The oil industry at large was shocked to learn that the Panama Case injunction had been granted. Attorneys for the Department of the Interior at once appealed the case to the United States Supreme Court. The immediate result of this ruling of the federal court was to give fresh impetus to the operations of producers and refiners who had been persistent in their defiance of both federal and state legislation.

The Panama Case resulted in the introduction in the Texas legislature, then in session, of a series of bills designed to give the Railroad Commission ample power to inspect refineries and refinery records, and to give to the Attorney General a clear definition of the powers of his department to prosecute violators of the orders of the Railroad Commission.

Later, as an additional aid to clearing up the situation, an amendment to the General Deficiency Act was passed by the United States Congress, imposing a tax on all oil produced in the country, the intent of this provision being not so much in the nature of a tax as to give the federal authorities the right to inspect the books of producers and refiners and thus ascertain whether the crude they handle is legally produced. This legislation has resulted in a noticeable reduction in the amount of illegally produced oil in East Texas, Conroe and Oklahoma City fields, strengthening considerably the crude oil and refined products markets.

Coincident to the granting of the injunction by the federal court in the Panama Case, it became apparent to the state officials in Texas that additional legislation was urgently needed to increase the powers of the Railroad Commission. Accordingly, in February, 1934, a special session of the legislature was called in an effort to obtain the enactment of such laws.

House Bill 99, known as the Refinery Control Bill, was designed to compel all the refineries operating in the state to report the source of oil which they handle. During the investigations carried on during this special session, ample proof of the culpability of a large percentage of the East Texas refineries in dealing in illegal oil was furnished. Of the sixty plants in operation in the field at that time, only four were rendering the required reports of the Railroad Commission. The receiving capacities of the sixty plants was approximately 140,000 barrels of crude daily. Shipments of refined products, which could be checked through tank car movements, indicated that about sixty thousand barrels of crude was being refined. Scrutiny of the pipeline connections of all sixty plants revealed that the allowable production from the wells selling their output to the refineries was but 18,000 barrels daily. The inference from these figures is obvious.

House Bill 99 passed both houses after facing one of the most powerful oil lobbies ever witnessed in Austin. The independent refiners made a desperate fight to defeat the bill, but the lobbying efforts of the major companies were even more determined. Representatives of almost every major company operating in Texas were present in Austin pending the outcome of this legislation, and they remained active until victory was apparent.

After passing both houses, only the signature of the governor was required to make the bill effective at once. Governor Miriam A. Ferguson, after some deliberation, announced that she would withhold signing the bill until a hearing might be held before her from both the proponents and the opponents of the bill. Two days were set aside for the hearing. Special trains were run from the oil fields to the capitol and several hundred oilmen took their pleas to the governor.

Having viewed one side of the matter, the governor awaited the delegation of opponents to the measure. With the intent of carrying their arguments by dramatic appeal, a special train brought the left-wing forces into the capital. Scores of the complainants went dressed in overalls and rough field clothes, parading, for effect, through the streets of Austin before going before the governor. It was learned that in order to fill the special train on this occasion, some of the opponents of House Bill 99 financed the trip for any volunteers who wished to go but who did not have the price of the railway fare. The story goes that numerous men were picked up from the street corners of Kilgore in order to augment the force. The independent refiners, most of them owners of small topping and skimming plants, were anxious to carry their side of the matter by picturing to the governor and her staff the distressing conditions that would prevail in East Texas if the fifty-odd plants in the field should be forced to close up and throw several hundred men out of work.

After both sides had been heard, the governor signed the bill and it became at once a law. At the same time, other bills were passed in the special session increasing the pipeline tax from one-tenth to one-eighth of one cent per barrel in Texas, this additional income to be used to provide a larger enforcement body for the Railroad Commission. The governor also signed these bills.

In answer to the complaints of the independent refiners, several of the major companies, following an offer made by the Humble Oil & Refining Company, declared themselves in readiness to allocate five percent of their production in East Texas to the refineries in the field in order that these independent refiners might not be forced to shut down for want of legitimate crude supplies.

Considering the fact that the independent refiners have long been a thorn in the side of the refiners of lawfully produced crude, the companies offering to turn over a small part of their crude will be making no real sacrifice. These fifty or sixty small plants had been buying their crude at prices ranging from twenty-five to forty cents a barrel, or, say, an average price of thirty-three cents, whereas the posted price for the field at that time was one dollar a barrel. The independent refiner, after processing his thirty-three cent crude, obtained about fifteen gallons of third-grade gasoline which he sold at about three cents a gallon at the refinery. Utilizing the remainder of his barrel of crude, the refiner produced about twenty-four gallons of fuel oil, for which there is a sale at one and one-quarter cents a gallon, and recovered at that time a gross of seventy-five cents per barrel of crude. Assuming that his refining costs were ten cents a barrel, he had made a profit of thirty-two cents per barrel from his investment in illegally produced oil.

On an estimated total output of one million gallons of gasoline daily from the East Texas plants, it may readily be seen that the legitimate refiners were anxious to remove the high percentage of bootleg gasoline from their field. The argument is even more conclusive when it is considered that the gasoline produced from lawfully produced crude, purchased at one dollar the barrel, must cost the legitimate refiner, placing his product on the wholesale market, about four and one-half cents a gallon. Most major refiners had

264

to place a so-called "competitive" grade of gasoline on the market. This product was produced and sold at a loss of one and one-half cents per gallon in order to meet the competition from gasoline made from illegal crude.

After lengthy consideration, the association of independent refiners in East Texas finally announced that they would accept the offer of the major oil companies to allocate them five percent of their production. A number of the plants have been closed, but at least half of them are still in operation. Time alone can tell whether or not these numerous small plants can survive in competition with the more efficient, modern cracking plants of the major companies.

POLITICS:
REBUKE

Strange as it may seem, while the Railroad Commission of Texas has been engaged in the battle with violators of its proration and conservation rules, it has tarried three and a half years in attempting to prevent over-drilling of the East Texas field. At the time of the Joiner discovery, the spacing rules of the Commission permitted locations to be made 150 feet from property lines and 300 feet from other producing wells on the same tract of land. When the vast proportions of the new field became known, the rules were altered to apply to East Texas so that wells would not be located nearer than 330 feet to property dividing lines nor 660 feet from any other producing well.

Unfortunately, the guiding principle in the production of oil seems to be to steal your neighbor's oil from under his ground before he can steal yours. It is the custom of the industry, sanctioned by long usage, and in the language of Portia to Shylock, "The law allows it and the court awards it."

This phase of oil production and the reluctance of the industry to remedy it have caused the unnecessary, wasteful and financially disastrous orgies of over-drilling which have followed the discovery of every new oil pool. It has been, and is still, the root of almost all the industry's troubles.

The existence of legal sanction for the predatory characteristics of oil production is rooted in the archaic judicial assumption that oil and gas, being migratory by nature, were akin to wild game, becoming property only when reduced to possession. Given that background, human greed and rapacity have worked unhampered to reduce the industry to economic chaos. Not until gasoline sold for five cents a gallon at the filling stations in California and a barrel of crude went begging for purchasers at five cents in East Texas did the industry wake up to the fact that if oil and gas were wild game it might be advisable to enact some laws for protection.

Some of the petroleum states, notably California, Oklahoma and Kansas, prior to the East Texas epoch, had passed so-called conservation laws to prevent waste. But the measures proved ineffective because they were concerned with effects instead of causes, because they offered no protection against a rapacious neighbor, because, in short, they failed to provide shackles for the wild game hunters.

The law presumes, in the absence of legislative enactment to the contrary, that an oil producer has the right to drill up his property as fast as he can, regardless of the law of supply and demand, regardless of the damage he may inflict upon his neighbor or of the havoc he may wreak upon the industry at large by wasting a valuable natural resource.

Disregarding frank declarations from the federal courts that recognition of property rights in petroleum with correlative enactments to regulate and restrict drilling, were within the powers of the various state legislatures, the oil industry has preferred to follow a fruitless course of attempting to obtain a balance between supply and demand through voluntary agreements.

To the great army of investors, who in recent years have poured billions of new capital into the industry, this studied ignoring of the basic causes of overproduction has been perplexing as well as disappointing. The truth of the matter is that, although it has been evident for several years that the business was being carried away by its own momentum, it has been impossible to slow down the drive that had become habitual under an earlier economic pressure.

Proration, then, was not the sole solution of the problems of overproduction in East Texas. Legislative control of drilling should have been one of the first remedies applied to the control of East Texas, but it was not. Proration, at first, became something of a gentlemen's agreement, under which only a portion of the producers agreed to abide by the rules. The defects of the early proration laws soon were apparent. When the producer turned prorationist (in the expectation of getting a better price for his reduced output) he submitted by agreement to a restriction on his potential output, not on his actual output. At the same time he did not agree, nor was he asked to agree, to restrict his drilling activities.

The result was that given an allowable output based on his potential production, i.e. a figure representing the total number of barrels his well would make if opened wide open for twenty-four hours, all he had to do to double the daily output was to drill a second well, and so on so long as there was space left on his lease for additional wells.

Proration, therefore, instead of fostering conservation, encouraged unnecessary, wasteful drilling and, by creating an enormous unusable reserve of developed production, entirely reversed its economic object of maintaining prices.

In an article entitled "Cooperation: The Way to Recovery" in the January, 1934, quarterly issue of the American Petroleum Institute, Axtell J. Byles, president, remarks:

The public, as well as the oil industry understands that overproduction of crude oil is responsible for most of its difficulties and the economic waste and evil practices which have followed. It may not be as well understood, however, that the industry up to this time has been unable to achieve a balance of supply with consumer demand by reason of a combination of the law of capture, the laws against restraint of trade, and of small minorities within the industry which—either by reasons of avarice or financial exigency, or both—have refused to cooperate in any constructive program.

Under federal control, which commenced September 8, 1933, daily production of crude oil has been reduced, with the cooperation of the oil producing states, from its then level of 2,721,000 barrels to 2,290,000 barrels as of December 23. In addition, it is estimated that about 60,000 barrels a day of "hot" or illegal oil are still being produced. While daily crude oil production has exceeded federal allowable, with the exception of the month of November and the first week in December, substantial progress has been, and is being, made in the right direction. Earnest and intelligent efforts are being exerted to bring daily production within the allowables and entirely to eliminate illegal or "hot" oil. It is recognized that the industry cannot attain a sound position so long as "hot" oil is produced.

Other than control of crude oil is necessary to stabilization pending the time when excess stocks of crude and products are liquidated, viz., runs to stills and correct handling of the present excessive stocks of products as well as crude.

No plan of stabilization can be effected unless it is recognized that every pool and operation in each section of the country has a responsibility to every other section and operation. The claims of different sections of the country from time to time that their supply and demand are in balance and that price wars do, nevertheless, occur will not bear close analysis. If quotas are set too high; if they are based on the old fallacy of refiner demand, which is speculatively influenced, instead of upon consumer requirements for products; if quotas are inequitably and uneconomically allocated between states and pools, if they are not respected and enforced—chaos, cut-throat competition, waste and bankruptcies are inevitable.

Proration of crude oil is sound in theory and under present conditions, essential in practice. It has heretofore failed because it has been neither equitably nor scientifically applied, nor has it been enforced. Local or temporary balances of situations are of no avail when there is no confidence in their generality or permanency. Once the threat of failure is removed, the whole psychology changes. Confidence replaces uncertainty and fear. Whereupon our proven reserves and great potentials—vast for the needs of today but meager for the needs of future years—become stabilizing and reassuring factors, instead of monstrous things threatening to destroy an industry while they are destroying themselves.

It is conceivable that in an economy where inventive genius and technical skill have demonstrated their ability to outrun consumption with production, the government may for a time have to regulate production of many commodities—increasing the allowables when prices become too high, and decreasing them when prices become too low. Such regulation, if it comes, may continue until we are able properly to adjust the use of credit, the flow of capital, attain more equitable distribution of buying power, expand foreign markets, and through trade agreements avoid destructive competition.

By reason of the fugitive nature of crude oil in place, and the fact that without restraint by government or enforcible agreement the drilling of a discovery well usually leads to the immediate complete development and early exhaustion of a pool, I would hazard the opinion that in this industry federal regulation of crude oil production may be necessary for some time to come. Such a policy need not, and should not, involve governmental operation of the industry.

If we produce enough crude oil to supply consumer demand for products, less enough to permit modest withdrawals from storage until they reach economic levels, it will not be possible to overproduce gasoline. Free interchange of crude and gasoline between those who are long and those who are short of these commodities will result, inventories will be liquidated, and the oil industry will be prosperous. Confidence, once established in the stability of economic quotas of crude oil, willing buyers at fair prices will appear for both crude and products. The government can, by fiat, establish a price; but no government, by fiat, can establish a value or maintain a price against a market value.

Under government control of production there is no danger of runaway prices for motor fuel or refined products. Should such prices threaten, the remedy is in the hands of the government through its control of production.

As further evidence of the trend of thought of leading men in the industry, we give below the resolution of the American Petroleum Institute, adopted during the Twelfth Annual Meeting at Chicago, during November, 1931:

WHEREAS, The prevailing practices in the development of oil and gas pools and the production therefrom based upon the capture by each owner in the pool of as much as he can, as fast as he can, regardless of injury to his neighbor, has been proven to be wasteful of these irreplaceable natural resources, inequitable as between the owners of the common reservoir, and demoralizing to the whole petroleum industry,

NOW, THEREFORE, BE IT RESOLVED, by the Board of Directors of the American Petroleum Institute; That it endorses, and believes the petroleum industry likewise endorses, the principle that each owner of the surface is entitled only to his equitable and ratable share of the recoverable oil and gas energy in the common pool in the proportion which the recoverable reserves underlying his land bear to the recoverable reserves in the pool.

If we are to believe in the endorsement of The American Petroleum Institute, then it is reasonable to assume that the owner of a ten acre lease in a common pool of oil would be entitled to but one-tenth as much as the owner of a one hundred acre lease. In practice in the East Texas field, the owner of a one-acre lease enjoys all the advantages that accrue to the owner of a ten acre lease. He is permitted to have a well on his one acre and from the well to produce the same allowable as his neighbor produces. Because of the irregular shapes of properties in East Texas, where titles derive from the Mexican land grants of a century ago, because of the frequent finding of ancient surveying errors and because of reputed vicious trafficking in permits, it is not uncommon to find three or four wells drilled on one acre of land. The record in density of drilling, perhaps, is found in one instance where three wells have been drilled on one-half an acre.

An attempt was made by the Railroad Commission, as soon as the proportions of the East Texas pool were known, to fix the limit of drilling at one well for each twenty acres— a rational plan of development in most areas, but one which has not been adhered to in this one. The principal objection to the plan was that it appeared discriminatory to the owners of large leases. Far the greater number of leases in the field are in small tracts, and it was found that granting the owner of a one-acre lease the right to drill one well would be inequitable in comparison with the owner of a twenty-acre lease who had had to pay twenty times as much for the leased premises and still would have no more production than his neighbor with one acre. This attempt, therefore, met with considerable opposition and the Railroad Commission sought to include other factors in the plan for granting permits and regulating the allowable oil production.

One of the Commission engineers, E. O. Buck determined on a plan that would have provided a satisfactory method of fixing the well allowables on an equitable basis, considering the factor of acreage. The plan was to allocate the daily allowable, then 325,000 barrels for East Texas, by giving two-thirds of the allowable figure equally to all the wells in the field, with the remaining one-third divided according to the acreage in the lease and the bottom-hole pressure. The Buck plan was put into operation, but the resultant cry of unfairness, on the part of the operators of the smaller tracts, soon led to its discard.

For some time thereafter, operators of large leases were held to one well to each twenty acres, while owners of the smaller tracts were permitted to have one well regardless of the size of their property. It so happens that most of the small tracts in the field are owned by smaller, independent operators, desirous of obtaining the maximum number of wells possible on their leases, regardless of the number of offset wells their activities would entail.

As if there were not a sufficient number of wells in prospect under the then existing spacing rule applicable to the East Texas area the Railroad Commission of Texas issued an order dated June 13, 1933, in which it is stated that from evidence "adduced at the East Texas hearing held in Austin on June 12, 1933, that Rule 37 limiting the spacing of wells in the East Texas field, including Upshur, Smith, Rusk, Gregg and Cherokee Counties, Texas, should be amended in order that properties might be more fully developed so as to insure a maximum oil recovery and to allow owners of various properties in said field to protect themselves against inequities which might result from a strict enforcement of said Rule 37."

The amendment to Rule 37 is as follows:

RULE I: (a) Rule 37, adopted November 26, 1919, is hereby amended insofar as it applies to the East Texas Field so as to hereafter read as follows: "No well shall hereafter be drilled for oil or gas at any point less than six hundred and sixty (660) feet from any drilling or completed well; and no well shall hereafter be drilled for oil or gas at any point less than three hundred and thirty (330) feet from any property or division line; provided, however, the Commission in order to prevent waste, or to protect vested rights, or to protect any property against undue drainage by reason of the operation of the wells of any other operator, will, after hearing, grant exceptions permitting drilling within a less or shorter distance than hereinabove prescribed, upon application duly filed fully stating the facts, notice of such application and hearing having been first given to all adjacent lessees affected thereby; provided, that if all adjacent lessees affected thereby waive in writing notice of hearing on or objection to the granting of said application, the Commission may proceed to determine such application without hearing; and provided further that in cases of forced offsets the Commission may grant exceptions without waivers of hearing when it is evident that the wells desired are necessary to protect the properties on which it is proposed to drill them.

Unfortunately, the facilities of getting permits to drill additional wells on the smaller tracts in the field were considerably enhanced by the latter portion of the above quoted order. When all parties were in agreement on the spacing of wells nearer property lines than allowed under Rule 37, then waivers of the necessity of holding a hearing could be exchanged and the permits could be obtained at once from the Commission. Owners of adjoining leases therefore, in collusion with one another, could arrange for the drilling of numerous wells by mutual agrement, and thus boost their allowable production. Such arrangements always were at the expense of other property owners in the vicinity of their leases, as additional drainage from the common reservoir of oil in the ground resulted from the drilling of each new well; each new well tends to reduce the bottom-hole pressure, the pressure which causes the wells to flow, and hastens the time when all the wells in the field will have to be pumped.

Many lease owners, in order to increase the number of wells on their leases, formed dummy corporations to whom they sold off a portion of the acreage, to enable the new corporation to apply for permits to drill wells that, under the spacing rule, would not have been permitted. In fairness to the Commission, if the supervisor in charge of hearing could ascertain that a subdivision had been made for the sole purpose of obtaining additional wells it was usual for him to deny the permit.

In June, 1933, at the time the order amending Rule 37 was issued, there were above ten thousand completed wells in the East Texas field. Fresh impetus was given to drilling, due to the exchanging of waivers that permitted the Commission to grant permits for wells without hearing.

In August, 1933, the following order was issued by the Railroad Commission, without hearing:

IT IS HEREBY ORDERED BY THE RAILROAD COMMISSION OF TEXAS that no permit to drill an oil well shall be allowed until notice of hearing has been given and the hearing held by the Commission.

In cases where all offset or adjoining lease owners waive protest in writing and it clearly appears that such well should be granted as an offset well to prevent drainage, the permit may be granted without a hearing, provided that to meet drainage of direct, equidistant offsets, the Commission may at its discretion grant permits without notice and hearing or waivers.

A variation of ten (10) feet in distance from property lines of offset wells will be considered as "equidistant" under this rule.

This ruling, commonly known as the "Equidistant Offset Rule," was equivalent to throwing additional fuel on the flame. Drilling permits were now being granted without hearing, without waivers and without notice to the offset property owners. Of course, it was not intended to speed up drilling, but it had that effect. It was obviously possible, in many cases, to present an application for permit, stating that the desired well would be a direct equidistant offset to some other well of an adjacent property owner, when such was not the case. But as no hearing was held the facts in such cases could not always be ascertained until the well had been drilled and the damage done.

The crime of drilling wells without legal permits was not regarded as a crime by certain operators who drilled in the face of legitimate protests from adjacent property owners and accepted the five hundred dollar fine imposed by the State as the easiest method of getting the added production. Occasionally an irate property owner would file an injunction suit to prevent the drilling near his property line of a well that was obviously unjust. In the end, however, the well would usually be completed and, after the offender had paid his nominal fine for contempt of court, the matter would be ended.

On May 28, 1934, the Commission executed an abrupt about-face with regard to the granting of well permits. Rule 37 was amended and the "Equidistant Offset Rule" of August 30, 1933, revoked. Apparently the Commission had been receiving too much criticism for the latter. The present amendment to Rule 37 requires that no well may be drilled nearer than 330 feet to any property line, nor nearer than 660 feet to any other drilling or completed well, not even in the case of direct and equidistant offsets, without submitting the application to a hearing in Austin after the usual ten day notice has been issued.

In addition to the above conditions, the applicant for such well must submit a detail map, drawn to the scale of one inch equal to four hundred feet, showing not only his property but the property of each adjacent lease owner, including all wells, drilling wells or permitted wells accurately described. In addition, the map must be accompanied by a sworn affidavit as to the correctness of all of the information shown thereon.

With approximately fourteen thousand completed wells in East Texas on June 15, 1934, it is evident that the stringent amendment to Rule 37, limiting the spacing of wells in the East Texas field, comes about three years too late. It has come, however, in time to prevent the possible drilling of some six to eight thousand additional wells that might have been obtained under previous rulings.

Actual figures on the drilled status of the East Texas field show that in June, 1934, there was a well to each nine acres. Over-drilling has prevailed to a greater extent in East Texas than in any other large oil field in this country. There seems no great change in the drilling policies of all the operators in the field, large or small. The usual race to get his neighbor's oil before his neighbor gets his remains the guiding policy of nearly every operator today.

Dr. Frederic H. Lahee, while President of the American Association of Petroleum Geologists, in a paper on "The Occurrence of Oil and Gas" states:

> With a vast store of information derived from tens of thousands of wells, we know that both oil and gas, even though they are in the small pores of rocks, occur as isolated, definitely bounded accumulations, often referred to as "pools" or "fields." We also know that within each such pool the oil and gas are under pressure due partly to the weight of the overlying rocks and partly to hydrostatic pressure from the water flanking the oil within the porous reservoir rock. And finally we know that the gas, while under sufficient pressure, serves as a most valuable agent in moving the oil through the pores of its containing rock to the lower end of the well, and then in lifting this oil through the well to the surface. In this way the gas, as natural energy within the reservoir, is of very great importance in facilitating the extraction of oil from below ground.
>
> Any method of development—and by "development" we mean drilling the wells and producing the oil—which dissipates the reservoir energy, by wasting the gas, or taking the gas out of the pool unequally, or taking it out too rapidly, reduces the efficiency of the energy available in the reservoir and thus reduces the quantity of oil which will eventually be recovered. Wide experience in many regions has shown that evenly spaced wells throughout the area of the pool are more conducive to maintenance of reservoir pressure than uneven spacing where the wells are crowded together in some places and wide apart in others;

and experience has also shown that the decline in reservoir pressure which is generally observed in producing fields can be retarded and kept more uniform by regulating the flow of oil in all the wells instead of allowing them to produce wide open. A gusher, pouring out hundreds or thousands of barrels per day, is a thrilling sight, particularly for the owner, but as long as it flows unbridled and without consideration of its relation to other wells in the field, it is trespassing on the prospects and the rights of other owners in the pool, both owners of leases and owners of royalty alike.

Unless a pool is drilled and operated systematically and efficiently, with full regard for the proper utilization of the reservoir energy, not only are the individual rights of the owners jeopardized, but—more important—oil in large quantities is left underground where it can never be recovered. No matter how carefully the wells are drilled and operated, there is always a certain percentage of the oil which cannot be extracted, and which is therefore left in the pores of the reservoir rock; but by poor practices in drilling and operating, a much larger proportion of the oil originally present in the pool may be permanently left in the ground.

What a pity it is that the sage counsel of Mr. Byles, Dr. Lahee and other learned men of the industry, has not been incorporated in the legislation enacted in the various oil producing states. Most of such legislation reminds us of "locking the stable after the horse has been stolen," but it is hoped that when future oil fields are brought into existence, the evils of the present will have brought about

281

better regulatory measures looking toward truly scientific conservation without economic waste.

With growing sentiment in favor of securing the passage of laws in the national congress to place the federal government in position to bring *real* enforcement of proration and conservation in the various flush fields of the country—especially in the prolific East Texas pool—lobbying in Washington has been actively pushed by interested oilmen and organized bodies representing operators of all classes, big and little. The Railroad Commission of Texas has jealously guarded its precious prerogatives, fearful that the leading industry of the Lone Star State might become hopelessly entangled with federal policies. Commissioner E. O. Thompson, elected to succeed former Commissioner Neff who resigned the office, has spent many weeks in Washington to see that the lobbyists' efforts to secure absolute federal control over Texas were defeated. Commissioner Thompson voiced the opinion that the Commission could and would secure complete enforcement with its own body of men.

In answer to insistent demands from the oil industry itself that something by done about the "hot" oil situation in East Texas, the Railroad Commission removed from Austin one of its most able executives, Chief Supervisor R. D. Parker, and despatched him to East Texas headquarters as Chief Administrative Officer of the Oil & Gas Division, effective April 1, 1934. The Commission, in an order dated Austin, Texas, March 31, 1934, signed by Commissioners Lon A. Smith and Ernest O. Thompson, authorized Parker "to organize said Division according to his best judgment, on as effective working basis as possible with the resources at his command, and to that end he is hereby authorized to employ such additional men of his own selection who are qualified to perform the services for which they are employed, experience being considered as necessary for the administration of said laws and to release from the service of the Commission employees now serving said Oil & Gas Division, if in his judgment, they are not qualified to perform the duties of the position they now hold…"

With this announcement, the industry in East Texas felt some appreciable changes in the regulatory body might take place. Mr. Parker fearlessly and energetically set about to put the East Texas house of the Railroad Commission in good order. He hired. He fired. He labored long and arduously to bring about an improvement in the enforcement of proration and checking of violations. So great is his popularity with the right-wing of the petroleum industry, many of his friends have repeatedly urged him to enter the state primary for election to the Railroad Commission to fill the place of Chairman Lon A. Smith, whose term will expire in 1934. Parker refused to have his name added to the list of candidates in the primary, preferring to carry out the stupendous task of curbing violations in East Texas.

Evidently Parker encountered numerous obstacles in his efforts to revamp the East Texas division of the Oil & Gas Division. Men he had unconditionally fired were rehired by the Commission and placed back in their old positions. Feeling that such a situation had become untenable, Parker tendered his resignation and left the services of the Railroad Commission on June 18th, 1934. His letter, made public through the press, is as follows:

When I was sent into the East Texas oil field, April 1, with a mandate from the Commission to stop hot oil production,

*I was pledged complete cooperation and promised that I
would be cloaked with full authority to hire and fire such
of the personnel as would be necessary in my judgment to
effectively cope with the then prevailing disgraceful situation, and in addition thereto was authorized and directed
to place the Oil and Gas Division of the Commission on
an efficient working basis with the admonition that efficiency was a prerequisite to the proper administration of
that division. It is well known by everyone that it broke
faith by breaking both pledges within two weeks after I
took charge of the East Texas field.*

*Apparently, I made the fatal mistake of firing political
favorites of all three Commissioners, some of whom had
been on the payroll of the Commission for several years,
although their records of achievement and efficiency have
been confined almost exclusively to vote-getting activity
during political campaigns. Every time I fired one, some
one of the three Commissioners would put him back at
work, with resultant weakening of the morale of the entire
organization. Moreover, their reinstatement resulted in
the use of money for their salaries and expenses, which was
then and is now sorely needed for employing investigators
to stop the illegal traffic of oil in the great East Texas field.*

*The first move to break the morale and shake the confidence
of those serving with me was when the chief supervisorship
of the Commission's Oil & Gas Division was taken from
me. However, in my zeal to get the job done in East Texas,
I swallowed my pride, redoubling my efforts and saying
nothing.*

*Therefore, forgetting my own personal welfare and real-
izing the utter futility of going further, I gladly sacrificed*

myself on the altar of stability of the oil industry by tender-
ing my resignation, with the hope that the three members
of the Railroad Commission speedily may realize that their
office carries with it a grave responsibility and is not a
concession conferred on them by the electorate of Texas to
be used as a means to further their own political fortunes
and to reward their personal and political friends with
positions and abortive favors.

I never was a yes man, and I never will be, and now that it
is definitely determined that this is what the Commission
wants, I step out of the picture and telegraph advance sym-
pathy to my successor if he assumes his duties determined
to stop oil thievery in East Texas, for it cannot be done
with the Commission's politically constructed enforcement
ideas.

The Commission, however, took the view that they had
discharged Mr. Parker from service, the dismissal being in
the form of an order signed by Commissioners C. V. Terrell
and Ernest O. Thompson. (It did not bear the signature
of Lon A. Smith, a candidate for re-election.) The Terrell-
Thompson order stated:

R. D. Parker, with a free hand and unhampered by any-
one, has had complete control of enforcing the oil laws in
East Texas for the last two and one-half months.

The production of excess oil steadily has increased to an
estimated amount of around 100,000 barrels a day since
he has had the sole direction of the enforcement of the law
and our orders pertaining to that field.

It is therefore ordered by the Texas Railroad Commission
that the services of R. D. Parker are hereby terminated…

Thus ended the career of a faithful and efficient employee of the Texas Railroad Commission who entered its service on January 1, 1909. The reader may draw his own conclusion!

PANORAMA:
ROYALTIES

The customary division of the oil lying under the soil is seven-eighths to the purchaser of the lease and one-eighth, called the "royalty interest," retained by the landowner. The burden of drilling, equipping and operating is placed upon the owner of the seven-eighths lease, and his interest is called the "working interest." Rightfully so, as all the work of development goes with this interest. The royalty interest is not assessable with any of the costs of getting the oil above ground.

This division of the oil results in the royalty owner being the aristocrat of the industry—provided that he has retained his royalty interest in the land. As a rule, nearly all landowners have parted with some of their royalty before oil was produced upon their land. There is a strong inducement to the landowner to dispose of some of his royalty and have the cash to show for it in case the land is unproductive of oil.

Before the discovery oil in a wildcat territory, royalties may usually be purchased for a little more than the amount per acre obtained in selling the leasehold interest. Where wildcat leases sell for one or two dollars an acre, the price of the royalty interest under the same land might be two to five dollars an acre. An acre of royalty is the full one-eighth interest under one acre of land. In some cases, where wildcat acreage is being blocked up preparatory to drilling, the landowners are induced to "throw in" a substantial part of their royalty along with the lease. This is a practice that is unfair to the landowner, and one considered unethical in the industry, for, as a rule, the landowner does not fully appreciate the value of the royalty interest.

Royalties under proven oil producing lands are spoken of as "producing royalties." Their value in the East Texas field today depends, of course, upon their location in the field. On the western side of the entire field, the pool of oil is flanked by water, and the water drive is constantly moving eastward as the oil is removed from the reservoir. On the eastern side of the field, called the shoreline, the sand pinches out until it is so thin as to be unproductive. In between the two shorelines, there lies the choicest production—the sand is thicker, bottom-hole pressure is greater and the ultimate recovery of oil promises to be greater.

Choice royalties, therefore, may run from one thousand to fifteen hundred dollars an acre for the full one-eighth interest. The price varies with the location, thickness of sand, as stated, and also varies in accordance with the number of producing wells and the reputation and ability of the lease operator. Figures may be obtained from reliable geological firms to show the estimated possible total recovery of oil per acre in any given part of the field. Owners of royalty, in preparing their income tax valuations on royalty, are permitted to establish some figure to represent the amount of oil their royalty holding will produce in the future, and thus arrive at the value of the interest. Naturally, the price of oil, based on the estimated recovery, is a purely theoretical figure and it is impossible to prescribe any rule for these calculations that will be entirely accurate.

Every oil field produces some material for the "rags to riches" theme so popular with the readers of Sunday supplements. Undoubtedly, East Texas, owing to its great oil field, will contribute much to such lore. One has only to look about the new oil cities in East Texas to appreciate the fact that oil is pouring millions of dollars into the coffers of the natives. There is a single lease in the field with one hundred and thirty wells completed and producing on it, owned by the largest major company operating in East Texas. The royalty paid to the landowner amounts to more than fifteen thousand dollars monthly. There are probably not more than fifty landowners in the entire field whose royalty incomes will approach this figure.

One landowner decided to dispose of his entire royalty to a major oil company on the basis of five hundred dollars an acre. The transaction was made at a time when five hundred dollars was a very fair price, and the owner of the royalty received about a quarter of a million dollars in cash. Then, as fewer and fewer desirable royalties were put on the market, prices advanced to double the prices prevailing in 1932 and 1933. The same royalty would bring approximately one thousand dollars an acre today. The unfortunate individual who sold this royalty was induced to invest his money in various promotions until he was very soon relieved of the entire fortune. The royalties, had he retained them, would no doubt have provided him with an income of forty to fifty thousand dollars annually.

A vast amount of promotion goes on in connection with royalty sales, and it may be well here to give some indications regarding the practice. Division orders of pipeline companies purchasing the oil reflect the names of royalty owners scattered over the four corners of the United States. One may wonder how the people living far from the activity in East Texas would find the opportunity of investing in royalties. The answer is that they acquired their holdings through brokerage houses organized in the principal cities of the country, with buyers stationed in the oil field.

A large percentage of these brokerage firms operate along perfectly legitimate lines, and, to small investors, royalty buying represents a means of obtaining a fair return on their investments without any considerable risk, except the fluctuation in the price of oil. East Texas royalties may continue to pay dividends to investors for the next fifteen to twenty years—the estimated life of the field. The royalty investor may consider his investment sound and conservative if it will return the capital investment within four to six years. Some of the royalty holding companies issue certificates of interest, and the income from royalty investments from many fields is pooled, the investor receiving dividends much the same as though his investment had been in common or preferred stock in some other enterprise.

Every new oil field is invaded by promoters whose sole object is to fleece the unsuspecting, gullible public. The sucker-lists in the hands of these promoters reveal that the majority of the investors are hard-working business and professional people—doctors, dentists, school teachers, ministers and, yes, even lawyers. The usual method is to send out glowing literature, adorned with pictures of oil wells flowing thousands of barrels of the precious fluid hourly, with descriptive matter that would undoubtedly shame the original Bonanzas of the forty-niners.

Testimonials of sudden rise from rags to riches, effusive geological data, etc., serve to kindle the enthusiasm of even the most sceptical prospect. Soon the money begins to pour in to the promoter, who, in order to save his own skin, must have some producing royalty with which to mulct the public. The transactions usually deal in twenty-five, fifty and one hundred dollar units in selected tracts of land, and there is no limit to the number of shares of interest the promoter may sell. So long as the money continues to pour into his pockets, the promoter may sell undivided interests in his royalty holdings.

From a list examined in the offices of one of the pipeline companies in East Texas, it was found that one promoter had obtained an interest in royalty under about one hundred acres of land. His interest was but 1/4096 part of all of the oil to be produced from this tract, yet it had been sold to more than forty investors in Canada. The instruments by which the shares in this royalty holding were transferred to the various investors included a stipulation that the seller would retain a full power of attorney from the purchaser, that he would receive payment for their interest, sell it if he should choose to do so, and remit the proceeds to them.

Under scrutiny we observe that forty-odd investors in this small fraction of royalty would receive one dollar out of each $4,096 of oil produced and sold from the hundred acre tract—provided always that the promoter decided to send it to them. The oil run statements show that their interest under four of the wells in his hundred acre farm amounts to approximately two cents per well per month. As there are about twenty wells on the tract, the forty investors receive forty cents monthly, or not much more than one cent apiece. At this rate, assuming each one had invested twenty-five dollars for his interest—it may have been much more—it will require twenty-five hundred months for each one to recover the amount of his investment. It is doubtful if the field will produce a sufficient length of time for these forty investors to recover their initial investment. The postage required to mail out the individual interest of each investor during the productive life of the field would exceed the value of the payments.

The United States postal authorities are ever on the lookout for these get-rich-quick mail promotions, and the suckers may at least have the satisfaction of knowing that some half-dozen of the promoters from East Texas have served a stretch in federal penal institutions. On the other hand, due to the vast area covered by the oil field, a few promoters had the unexpected good fortune of keeping out of the hands of the authorities when royalties, at first worthless, became productive.

One of the semi-major oil companies, with important holdings in East Texas, allowed the employees of the company to form a royalty buying pool, in which ten to twenty percent of their monthly paychecks might be placed. One of the executives of the company was designated as buyer of royalties; all the profits to go to the employee-investors. To date, the employees have paid in to the pool a total of approximately twelve thousand dollars, and have drawn dividends from their investments amounting to forty-six thousand dollars. Properly managed and without the promoter or middle-man to take the lion's share of the profits, no saner method of investment might be conceived of for the salaried employees of the oil companies. It provides them with an incentive to save and earn, builds up their respect for their employers and establishes an unquestionably superior morale than if they are forbidden to own any interest in royalties or leases in the fields—the usual major company policy throughout oil country.

As suggested in an earlier chapter, many of the original lease purchases were made for part cash and a balance payable out of oil. With so large a part of the capital investment of the field coming from small independent operators with limited finances, a great deal of the financing had to be done out of the operators' oil interest. These payments are known in oil country as "oil payments" and provide that out of a fixed portion of the operators' seven-eighths interest, a certain sum is to be paid to the holder of the oil payment.

Drilling contractors, supply houses furnishing the material to the producers, and others having dealings with the producers, have come to recognize the oil payment method as an important part of the financing of new development. Bankers in East Texas, as well as in other parts of the country where oil is a by-word, have learned that the purchase of oil payments is very profitable business for them. The payments may be discounted for cash at numerous banks in the oil belt, the customary discount being about fifty to sixty percent. Hence, in order to finance the drilling and equipping of a well, the producer may raise the cash at one of the banks by assigning an oil payment for about two to two and one-half times the amount of cash needed for this purpose. An oil payment in the amount of thirty thousand dollars may be required to raise the twelve to fifteen thousand dollars cash needed. Everywhere in East Texas oil country, fine homes, cars, business property and other valuables are offered in exchange for choice oil payments.

How much of the money paid for East Texas oil stays at home? Of the 500,000 barrels produced daily and marketed at one dollar a barrel, one-eighth represents royalty oil. Royalty owners, therefore, probably receive about $62,500 a day from their East Texas holdings. The landowners themselves doubtless have retained about one-half of this royalty. Thus, native East Texans, whose ancestors pioneered the country in the days of the Republic, are receiving $31,250 daily, or eleven to twelve million annually. This calculation does not take into consideration the payrolls of operating companies, refineries, gasoline plants, supply houses, machine shops and foundries, ninety percent of which is doubtless spent in the oil communities for food, clothing, shelter and recreation.

The towns bordering on the oil fiend—Henderson, Overton, Longview, Kilgore, Gladewater, Arp and Tyler—have experienced rapid growth, populations having approximately doubled since the discovery of oil. New buildings of all descriptions have been erected—skyscrapers, churches, schools and homes. Two out of every three passenger automobiles appear to be new. There is plenty of money in East Texas and most of it is kept in circulation.

PANORAMA:
East Texas & the Future

If the geologists are right, East Texas will produce an ultimate total of two billion barrels of oil. On the basis of 120,000 acres of proven producing leases, the result will be the average recovery of 16-17,000 barrels of oil for every acre. Such being the case, East Texas has already utilized one-fourth of its potential production.

If every barrel of the five hundred million barrels already produced and marketed had sold for one dollar, which it has not, then a total of half a billion dollars has been returned to the producers and royalty owners. If we pause to consider the investment in oil properties in this field, it will be apparent that first costs, which include the amounts paid the landowners for their leases, drilling and equipping the wells, laying pipelines for the transportation of the oil, building refineries, plus the labor bills incident to the development during this 3-1/2 years, will total approximately one billion dollars. An additional five hundred million barrels of dollar oil must therefore be recovered before investments will have paid out. This is equivalent to saying that one-half of the potential recovery of oil from the East

Texas pool will go to pay for development. Out of the remainder, the producers must look for their profits.

It is safe to predict that East Texas will have twenty thousand wells before drilling ceases.

About 660 of the more than 13,000 wells completed on June 1, 1934, were pumping wells. Most of them are located along the eastern shoreline of the field. The cost of installation of pumping equipment varies considerably with the type of machinery employed. The average cost may be in the neighborhood of $3,500. Thus, an additional sixty million dollar investment will be required before the entire field is equipped to pump—encouraging news for the companies supplying this type of equipment to the industry.

This phase of production will call for the employment of many additional men. Naturally, the pumping well requires considerably more attention than does the flowing well. The time that may elapse before all the wells will have to be pumped has been variously estimated to be from eighteen months to three years from the present. It is likely that the latter estimate is more nearly correct. Reservoir energy, the combination of rock pressure on the oil producing sand and the hydrostatic pressure, must fall off considerably before the wells cease to flow. Early in 1934, the average bottom-hole pressure was 1,264 pounds to the cubic inch. This pressure must decline to somewhere around 980 pounds before it will be insufficient to bring the oil to the surface. There was an average monthly decline of ten pounds of pressure during the early months of 1934. If the field can be kept under control so that all the wells produce their allowable production ratably, there is little to fear in the way of early destruction of the reservoir energy that causes the wells to flow.

This raises the question as to whether or not proration is here to stay. Undoubtedly it is. Everyone who has experienced the orgy of overproduction that followed the discovery of this field is aware of the importance of continuing proration. Doubtless, if the field had been allowed to continue producing unhampered by governmental control, there would be little to show for the billion dollar investment in East Texas today. The wells would all have ceased flowing many months ago, the market would have become bankrupt, with rioting, disorder and social upheaval following in the wake.

There has been much discussion of the question of the State of Texas creating an Oil & Gas Commission to relieve the Railroad Commission of this duty. Bills proposing a separate commission have been introduced in recent sessions of the state legislature, but friends of the present commission have outnumbered its opponents.

Texas is the leading oil state of the union, producing during 1933 forty-four percent of the nation's oil, equivalent to one-third of the world's production. Occupying, as it does, this leading position in the industry, its oil and gas commission should be composed of men whose knowledge of the oil business has been acquired by actual experience in all phases of the work of drilling, producing, shipping, refining and marketing oil. Obviously there are few politicians capable of meeting these requirements.

Besides its duties in connection with the oil and gas industry, the Railroad Commission of Texas must deal with the affairs of transportation, railways, bus lines and trucks, with numerous public hearings to be conducted incident thereto. For this reason, if for no other, it is argued, the state might well create a separate Oil & Gas Commission, whose duties could be confined to the state's largest industry.

While the East Texas oil field is confined to portions of Smith, Rusk, Upshur, Gregg and Cherokee counties, it is apparent that other portions of the eastern part of Texas are due to become oil producing centers. The Van field, in Van Zandt County, has been producing since the latter part of 1929.

In the autumn of 1933, the Tide Water Oil Company and Texas-Seaboard Oil Company succeeded in completing a producing oil well on their five-thousand acre Long Lake Plantation block in the southwestern part of Anderson County. This discovery quickly led to frantic leasing of all the adjacent unleased lands. While it is evident that a field lies under the southwestern part of the county, the second and third wells to be drilled on the Long Lake Plantation were completed as gas wells making a limited amount of distillate. As a consequence, drilling activity was checked in that area.

The same operators who made a strike with their Long Lake producer moved to a block of leases in the northwestern corner of Anderson County, known as the Cayuga area, where they began drilling another wildcat well. Here again, on the J. N. Edens 988.6 acre farm, a gusher was obtained in the prolific Woodbine sand horizon. The completion, on March 1, 1934, was sufficient to send the price of available leases in this area to a record peak. Production on June 1 had been extended several miles north and east by the completion of eight additional wells in the Cayuga field.

The Railroad Commission of Texas, after calling a hearing on the development of the Long Lake district, promulgated rules for the spacing, drilling and completion of wells in that pool. These rules will permit the drilling of only one well to each twenty acres, with no well to be nearer than 466 feet from any property dividing line or 933 feet from any other well. The same rules have been applied temporarily to the development of the Cayuga pool, and, after public hearing on the matter, will doubtless be adopted permanently.

In the little town of Rusk, county seat of Cherokee County, has recently come the greatest wildcat excitement to occur in East Texas since the days of the Joiner discovery. On June 1, 1934, a well was completed in the Woodbine sand at a depth of approximately 4,900 feet. The owners of the well, Wood & Young, leased a large block of land from the New Birmingham Development Company, and the discovery has resulted in the revival of an old hope for fame, until now almost forgotten. The New Birmingham Development Company dates back to 1888, when the company was formed for the purpose of exploiting rich deposits of iron ore. A salesman, A. B. Blevins of Birmingham, Alabama, journeying through Cherokee County, observed in surface conditions the possibility of developing the iron industry in this county to a point were it might well rival Alabama development.

Blevins succeeded in securing capital for the formation of a company to undertake the Cherokee County project, and 20,000 acres of land was acquired about two miles west of the town of Rusk. Eastern capitalists were induced to promise to finance the venture and a city, called New Birmingham, sprang up on the company's holdings. A fine hotel was erected, at a cost of $60,000. Brick houses were built. Schools, churches, a railway station and numerous dwellings were constructed.

The promoters of the enterprise shortly discovered that eastern capital could not be obtained to finance their undertaking, for the simple reason that the creation of a new iron development would interfere with the iron markets in which the capitalists were already interested. After exhausting the possibilities of financing the Cherokee County development in this country, the backers went to London. Here they were successful in getting the interest of a number of Englishmen who came over to investigate the proposition.

Unfortunately for the promoters of the New Birmingham development, Texas had just passed certain alien land laws that prevented the investment of foreign capital in the project. Although a regal dinner was staged at the fine new hotel in New Birmingham, with Texas Governor James Stephen Hogg as honor guest, all efforts of the promoters to secure a change in the laws were futile and the members of the English syndicate returned to England without investing their millions.

Gradually, the people of New Birmingham began drifting away, a disastrous fire destroyed a large part of the town and decay set in. By 1900, the New Birmingham excitement had been forgotten. The Southern Hotel, largest monument to the enterprise, remained standing until destroyed by fire in March, 1926. No vestige of the city remained in 1934 to tell the tale of its one-time grandeur.

Now, however, with the discovery of oil, which may be used for fuel to reduce the iron ore in preparation for smelting, and with known deposits of lime nearby to be used in fluxing, and with sufficient coal to provide the coke necessary in the smelting process, New Birmingham should again come into its own. Iron and oil may combine to make Cherokee County an area to rival the East Texas oil field in wealth.

PANORAMA:
TEXANA

Scattered through the oil field is a small percentage of farm land owned by negro farmers. Unfortunately, most of them fell prey to promoters or became tied up to unscrupulous lawyers to whom they went for advice. A few, however, succeeded in retaining their leases and royalties until fair prices were obtained.

In the earlier days of the field, many interesting facts were brought to light by oil company representatives in the quest of missing heirs whose signatures were necessary on oil and gas leases and on royalty transactions. One of the strangest cases was that of a missing negro girl who, for obvious reasons, must remain unnamed. Her story is told by a representative of one of the companies which sought to purchase some royalty from her:

We were after a three hundred acre lease in Gregg County, which, while not exactly proven, showed up very promisingly. The farm had been the property of an old negro farmer who died before oil was found. The old man's daughter married and lived on the place with him. After she had had her third child, she developed tuberculosis and the doctors said the only hope for her was removal to a higher and drier climate. After much discussion, it was settled that the woman, her husband and the two older children, a boy and a girl, would go to Arizona, leaving the baby boy with the grandfather.

Well, they moved west and the mother died within a year. The father and the two children, all of whom were of light enough color to pass as white, moved farther west and settled down in a medium-sized city where they *did* pass for white. The boy and girl finished high school, and the father, who had bought a home and worked at the carpenter trade, married a white woman. His children did not care for the stepmother, so the boy went to a larger city to find work and the girl went to one of the large universities in another state.

Before she had finished her first year at school, her father died and they shipped him back to Texas for burial. The stepmother did not accompany the body and doubtless never knew that he was not of white parentage. The girl left by train for the nearest point to the old homestead, arriving the evening before the funeral. She stopped at the leading hotel in this town and the following day hired a car to take her to the aunt's place where the burial services were to be held. "Oh, you know old Aunt Martha, do you?" the driver inquired.

"I ought to," the girl unthinkingly remarked, "since she is my father's sister." This confession was unfortunate for the girl, accustomed by now to pass as white, for she had committed an unforgivable sin in her native homeland. The driver put her out hastily on arriving at her aunt's house and hurried back to town to spread the news. A considerable stir of excitement resulted, as there had been high feeling between whites and blacks, with an outbreak of race-rioting in Longview just a few days earlier.

The girl found her old negro aunt and scores of former friends at the house preparing for the services. Before the services had been completed, word came that a mob was on the way to drive the girl out of the community. Frantically, her colored friends spirited her away, just in time to prevent serious trouble, and she left that night to return to the university, vowing she would never return to her former home in East Texas.

When she arrived at the school again, she suddenly determined to give up studying for fear that her secret might become known. She wrote the old aunt, who had now taken the baby brother to raise, that she was going to travel as companion to an invalid white woman. That was the last that relatives in East Texas heard from her.

Prior to leaving the university, she was requested by the stepmother to take for her share of the father's property one-third interest in the three hundred acre farm in Gregg County, with one-third each for the other two children, leaving the home and money in the bank as the stepmother's share of the property. Feeling that they were getting the worst of the bargain, the girl and her older brother reluctantly executed the papers making the division in that manner.

Some years pass and the girl had gone to a large western city where she worked as a waitress. In the meantime, she was married to a white man, but obtained a divorce after a year or two. She drifted to another city after this, and again found work as a waitress in a popular dining room. She became engaged to a man at about the same time oil was discovered in East Texas, but I doubt if she knew that the oil was anywhere near her land in Gregg County.

When our company began the search for her, we traced her to the university town where she had attended school by means of letters she had written to her old aunt. But there, the trail ended. A rival company had picked up the trail and found a girl they thought to be the one they were seeking, but the girl they had found denied any knowledge of the case. Still unconvinced, the rival company secured the old aunt from East Texas and took her west where they confronted the girl they had "discovered" with the old colored woman. Although a good many years had elapsed since the girl's flying visit to and tragic flight from East Texas, the old aunt was positive that the girl whose identity was questioned was not her niece. The trail was again cold.

I kept on with the case and eventually found her. I had been fortunate enough to secure a picture of her, taken while she attended the university, and carried it with me to help in the identification. I managed to get her to come to my table when I went to the dining room where she worked. When she saw the picture in my hand, she said, "For God's sake, where are you from?"

I made an appointment with her and found that she had bought some real estate in the city in which she was living, and had given the attorneys for this company an option to sell her East Texas land. They sold her oil rights the week before I found her, and obtained eight hundred dollars for the lease—which was worth about one hundred times that much—but she still had her royalty.

I offered her twenty-two thousand dollars for one-half of the royalty, but could not get her attorneys to release it, as they had given an option on it to another oil company, a rival of ours. After several weeks, they sold it for her for approximately the same figure our company offered to pay. She retained the rest of her royalty, which must pay her now about eight hundred to one thousand dollars a month. She asked me about her old auntie, her brothers and other relatives, but made me promise not to let any of them know where she was. I told her that her older brother was living in the same city where I had found her and that her "baby" brother was attending school in East Texas. His guardian, a respected attorney, had leased the younger brother's lands to a good advantage and he was obtaining a very large income from the royalty.

A prominent East Texas attorney tells of the search for a negro woman who had deserted her East Texas home and family, running away with a negro man ten or twelve years before the discovery of oil.

A client of mine was very anxious to buy the lease on a large tract of land owned by the husband of this woman, who ran away leaving him with the care of the five small children. Nothing was ever heard of the mother to indicate whether she were living or dead, although at the time she left the homestead, she was supposed to have gone over to Louisiana. The property was community property and, under Texas law, could not be leased with good title until the facts concerning the disappearance of the wife were known.

The father, with all the children, was brought down to my office to talk over the proposition of signing the lease, and I could see that he was very bitter over his wife's desertion. He is now an old man, with white hair, very tidy in appearance. It is apparent that he has been deeply grieved over the unfaithfulness of his wife. They left the office and I did not see them again for some weeks.

One day, my secretary told me one of the boys of that family wanted to see me urgently. I had him come in and tell me what had happened. It seems that he had made numerous inquiries in the hopes of finding his mother alive, and had learned that she had settled down in a village near Shreveport, Louisiana, years before. The boy borrowed some sort of old car from one of his friends, got a couple of dollars in his pocket and set out for that village. Arriving there, he picked up fresh clues and finally located his mother, but not until he had pawned his watch for another dollar or two to buy gasoline for his car. He had made his mother promise to come over to Texas and sign the necessary papers so the oil and gas lease could be sold.

The day set for the meeting, the father and the children arrived at my office before the mother appeared. I had the oldest boy take the father into a private office, as I was a little doubtful as to his reaction on seeing his erring wife again.

By and by, the elevator stopped and the mother appeared. The children, except the oldest boy, had all lined up outside the door leading into my office, but the mother passed them by without acknowledging them as her own. I talked with her a while. Finally the father came in with the oldest son, but he either could not or would not speak to the woman. She signed the papers, received a check for her part of the lease money and then went out to see the other children. She could scarcely believe they were her own, but eventually identified all of them by scars or other markings that told her they were really hers. She sobbed when she left.

When she had taken the elevator, the others gathered around the father and slowly led him away. The checks they carried away with them were large, but I gathered they were not large enough to erase the memory of the meeting with the woman they might have loved and called "mother."

An Iowa doctor purchased six or seven hundred acres of East Texas farmland many years ago for what was then a nominal sum. When oil was found on his land, he gave up his practice and moved back to East Texas. Today he has the royalty from about seventy-five wells on his land.

A speculator, desirous of buying some particular royalty in East Texas, was approached by a relative of the owner— who resided on a farm in another part of the state—who volunteered to help him acquire it at a low figure. They journeyed to the small village where the old farmer resided. Although he was totally blind, they found him plowing in his field. The mules, trained to such work, guided the plow almost as well as though their master had had his sight. The papers were already prepared and the old man signed away his royalty for a dollar and a half per acre, although it was worth nearly five hundred an acre at the time.

When he learned that he had been defrauded out of a fortune, the farmer brought suit against the purchaser. At the trial, the lawyer for the victimized farmer, taking him to the stand to testify, had placed a chair in front of the jury box in such a manner that the old gentleman would stumble over it on the way. The jury, thus appraised of his infirmity, awarded him the verdict.

Then there is the story of the landowner at Kilgore who had all his life been very poor. As the oil activity gradually approached his farm, he resisted all efforts of those who offered to lease it. Finally, when there was oil production on all sides of the land, he was induced to sell the lease, obtaining a cash payment of about $50,000. When the money had been deposited to his credit in a Kilgore bank, the farmer went in and obtained twenty dollars of it. He then went directly to a clothing store to purchase the thing he had coveted for years—a $16 Stetson hat. With his remaining four dollars, he bought food...four dollars worth of bananas! Another man, with his new wealth, bought an overcoat and a restaurant.

Dealers in luxuries, such as fine cars, diamonds, silverware, oriental rugs and costly oil paintings, have done a thriving business in the cities of East Texas. Architects and builders have profited from the building of new homes. Many of the landowners who acquired riches from their oil rights have remained on the old homesteads, satisfied with a new coat of paint for the house and barn, the building of some new fences and the acquisition of a new low-priced car. Others have erected mansions in the heart of the oil field, desirous of living in homes that will reflect their owners' wealth, while retaining the old environment.

Driving into a filling station for gas, the average customer does not appreciate the fact that for every gallon of gasoline purchased, there has doubtless been, somewhere in the oil fields where the crude petroleum was produced, a drop of blood spilled. Drilling wells is an operation filled with hazards; the rates charged by insurance underwriters for casualty insurance testify to this. Frequently a piece of machinery, a tool of some description or other heavy object topples from the upper part of the derrick and falls some sixty to one hundred feet to strike some workman a fatal blow. Boilers, fired furiously day and night in the push to "get the job done ahead of the other fellow," sometimes explode, spreading havoc to all those within range. Trucks laden with heavy oil field supplies frequently go out of control and slither into ditches. On the congested oil field highways, motor accidents are excessive, and the toll in lives is tremendous.

Early in May, 1931, a gusher in the Lathrop area of the East Texas field was being brought in. In this area of large pressures, the mud in the hole is thinned out and the wells are allowed to "clean themselves" of the mud and water; when clean oil appears, it is then turned into the tanks. In some unaccountable manner, this one well got out of control and soon was blowing wild over the top of the derrick. Hundreds of barrels an hour were being sprayed high into the air, the mist-like particles creating a black cloud of the highly volatile fluid. Men rushed in to attempt to stem this Niagara by installing the necessary valves on top of the casting. The fumes from the natural gas were asphyxiating, and working in them impossible for more than a few seconds of a stretch. Then it happened! A spark that may have been caused by the contact of steel with steel, or from a rock cast out of the hole striking a spark from the steel derrick, turned the cloud of oil into a blazing hell. Nine lives were snuffed out in the first moment of the burst of flame. For weeks, the fire burned out of control. The glare from the flame was visible at night for more than fifty miles. Experts finally were successful in extinguishing the fire by blasting it with terrific charges of dynamite.

Three men met death when the motor car in which they were driving passed through a valley in which a dense layer of natural gas had settled. A sudden backfire from the car created for them a mighty funeral pyre.

In a peaceful vale, skirting the paved highway between Longview and Kilgore, a tourist family put up their tent to pass the night. They were miles from the nearest oil well. Above them, on a nearby hill, however, lay an oil pipeline, carrying vast quantities of oil under high pressure. Sometime during the night, the pipeline burst and a flow of oil coursed down the hill toward the tent of the sleeping tourists. Suddenly, on reaching the smouldering embers of the campfire the on which the family had prepared their last meal, the modest tent became a charnel house. Four lives were blotted out.

Perhaps it is so in every great industry—Men must die; the wheels of progress must not rest.

PANORAMA:
JOINER

Before the end of 1930, Mr. Joiner knew that he had really achieved the dream of every wildcatter—to bring in a large oil field, to receive the financial reward that such a discovery should carry with it and to enjoy the addition of his name to the list of those who have been successful in the annals of the petroleum industry.

It was quite natural that numerous opportunities would be offered Mr. Joiner to sell his holdings. Many men, on reaching the Biblical age of three score and ten, are ready to lay aside the cares of active business or profession and indulge in some idleness or recreation. Not so this intrepid wildcatter!

The Joiner holdings were sold to H. L. Hunt of El Dorado, Arkansas, an independent operator who had already achieved some success in the oil business in that state and in Louisiana. The entire lot of leases held by Mr. Joiner, consisting of approximately 4,000 acres in Rusk County, was sold to Mr. Hunt for a consideration of about one million and a quarter dollars. Seventy-five percent of the Joiner holdings lay in unproven territory, so a large part of

the purchase price was made payable from future oil production and was contingent upon oil being produced from
the various properties. That the deal was advantageous and
bespeaks the foresight of the present owner of the leases is
amply evident. Approximately 3,000 acres of the original
purchase of leases lies within the producing area of the field.

The development of the properties, held in the name of
the Hunt Production Company, has entailed the drilling
of two hundred and thirty wells, and the maintaining of
a force of more than two hundred men in the field. This
company, through its fortuitous purchase from Joiner,
ranks thirteenth among the leading operating oil companies in the East Texas field, a position that many of the
major oil companies have not attained in the district.

Joiner, far from quitting on obtaining the opportunity
to do so, enlarged his offices in Dallas and began a fresh
and vigorous campaign of wildcatting. The three and a half
years from the time of bringing in the Bradford discovery
well have found him actively engaged every minute in the
drilling of a wildcat in some part of Texas. So far, this activity has not resulted in the discovery of any new oil field,
but, to Mr. Joiner, such activity is essential to his well being
and from it he derives much of the pleasure of living.

The discovery of oil in East Texas has resulted in a vast
amount of litigation: titles to property are constantly under
attack in the various courts in Texas, and not a little of Mr.
Joiner's time has been consumed in defending himself from
such attacks. Some of the litigation has been of no merit
whatsoever; some has proven costly to the discoverer of oil
in East Texas.

The town of Joinerville, created under boom town conditions, has slowly disintegrated. As the center of oil activity

moved to the west and to the north, the business and trade of the little city moved away with it. The telephone exchange was forced to close and calls are now handled through Henderson. Many of the temporary buildings of the early days of the excitement have been removed. Little remains of the glory of Joinerville. The town is dead!

So, as is the case of the promising village named after the discoverer of the greatest oil field in the world, the field itself soon will reach the apex of its development and decline will begin. A few more months, a year perhaps, and drilling of new wells will be practically at an end. One by one, the flowing wells will cease to flow, gas engines and power units will be installed, and pumping will become a necessity. The checks of royalty owners will begin to diminish and finally will be no more. As the level of oil in the reservoir under the field recedes, water encroachment will take its place and abandonment of properties must begin. In a few generations' time, the East Texas oil field may become but a memory, a legend to be added to other stories of East Texas, with its colorful past, glamorous present and brilliant future!

www.ingramcontent.com/pod-product-compliance
Lightning Source LLC
Chambersburg PA
CBHW020338100426

42812CB00029B/3176/J